The Already Dead

—

The Already Dead

The New Time of Politics, Culture, and Illness

—

ERIC CAZDYN

Duke University Press
Durham and London   2012

© 2012 Duke University Press

All rights reserved

Printed in the United States of
America on acid-free paper ∞

Designed by Heather Hensley

Typeset in Arno Pro by
Keystone Typesetting, Inc.

Library of Congress Cataloging-in-
Publication Data appear on the last
printed page of this book.

CONTENTS

—

ACKNOWLEDGMENTS

—

Modified portions of this volume originally appeared in different venues: *Prefix* ("The New Chronic"), *South Atlantic Quarterly* ("Crisis, Disaster, Revolution"), and *Modern Language Quarterly* ("Anti-anti: Utopia, Globalization, Jameson"). Thanks also to those colleagues who invited me to present lectures on parts of this work: Taipei (Chi-she Li), Bogotá (Gregory Lobo), Tokyo (Hara Kazuyuki), Zagreb (the folks at MAMA), Beijing (Wang Fengzhen), North Carolina (Phil Wegner of the Society for Utopian Studies), Antigonish (Nicholas Brown of the Marxist Literary Group), and Toronto (Rebecca Comay of the Literary Studies Program). I also want to thank my research assistant, Ryan Culpepper, as well as Peter Fitting, Martin Zeilinger, Andrew Johnson, two anonymous reviewers, and my editor at Duke, Courtney Berger, all of whom read through the completed manuscript and provided indispensable feedback. Thanks must also go to Christine Choi and Neal McTighe at Duke for their careful and thoughtful assistance. Regarding legal and medical issues, I thank Stanley Bush, Randolph Hahn, Jeffrey Lipton, and Jeff Siewerdsen. Others who have provided essential support for this book are Adrian Blackwell, Marcus Boon, Eric Chenaux, Lily Cho, Eva-Lynn Jagoe, Fredric Jameson, David Kersh, Renée Lear, Masao Miyoshi, Abigail Pugh, Gary Rodin, and Imre Szeman.

When asked whether he would see a new patient, a psychoanalyst wanted to know if it was an emergency. "No, the patient's relatively stable, functional, adjusted—not in crisis," the
referring doctor replied. "Oh no!" the analyst exclaimed. "In
that case I better see him right away."

What attracts me to this scenario is the short-circuiting of a
whole series of expectations and assumptions, not only about
psychoanalysis, but about crisis, time, illness, cure, capitalism,
life, death, and politics—the very problems around which this
book is organized. The lesson here is that if you really want to
understand a system and make significant change (and not just
manage symptoms), you must look away from what appears to
be the immediate crisis and toward the crisis that is at work
even when the system is functioning well. This crisis constitutes the system itself; the system cannot function without its
internal crisis. Psychoanalysis is less about digging up buried
treasures than about carefully inspecting the ground—a relentless listening to and intervention with the patient when he or
she is most functional and thus symptomatic (or most symptomatic and thus functional). Psychoanalysis, therefore, is not
a last-ditch effort at disaster relief. It might serve as an object
lesson to study why this project of psychoanalysis departs so
radically from the vast majority of cultural representations of

psychoanalysis, representations in television, film, not to mention many clinical case studies in which the "aha moments" are perfectly cued by close-ups, cloying music, or narrative momentum. *Crisis is not what happens when we go wrong; crisis is what happens when we go right.*

The logic of this last sentence will serve as a sort of mantra throughout this work. Take, for example, capitalism: crisis is not what happens when capitalism goes wrong, but when it goes right. In this light, Alan Greenspan's mea culpa in October 2008 (when the bankruptcies and bailouts began to flow) is particularly revealing. Greenspan was shocked; he thought he understood how things worked. But what really shocked Greenspan is that today's capitalism is still capitalism. He believed that we were in some mode of production that could have the productivity and wealth-generation of capitalism without capitalism's crises, or at least a capitalism in which the crises could be managed by mathematicians and derivative dealers. What a horror it must have been for Greenspan to realize that capitalism is still the system it has always been: one that cannot suspend its fundamental rules of profit creation and expansion under any circumstances (and thus cannot suspend class conflict, gender inequality, imperial violence, brutal dispossessions, ecological destruction, and psychological suffering). The point is that crisis is built right into the system of capitalism—not only when it busts, but when it booms.

Exploitation occurs within capitalism, therefore, not only when pensions disappear, jobs are slashed, and factories burn down, but it is also present when contracts are obeyed and factories are clean and safe. To prove this point, one should consider the valorization of surplus value and the logic of the commodity. For example, the inequality built into capitalism is due not simply to the pursuit of profit or so many bad capitalists, but to the seemingly natural and benign control a capitalist has over a worker's labor power. That labor power is reproduced daily with the making, selling, and buying of commodities—especially when the capitalist is kind, compassionate, and humane. Such an understanding of capitalism, one that emphasizes a larger structural logic over the personalities of the individual actors, is, in fact, the foundation of all economic theories, from neoclassical to Marxist. Moreover, understanding capitalism in such a way compels us not to moralize cheaply about factory layoffs or be incredulous about financial meltdowns. These should be expected by anyone who understands capitalism. In other words, these adverse effects are not simply about greedy politicians and corrupt executives. There are plenty of them. But to focus on them is a strategic error.

In March 2009, headlines were filled with outrage over the bonuses paid to Wall Street executives. The problem, however, was, and still is, not the bonuses at American International Group, but the radical inequality of income that sustains the larger economic system itself. Yes, outrage should be directed at the executives who received lavish bonuses despite selling the very financial products that exacerbated the meltdown. But a more focused and sustained outrage should be directed at the hard socioeconomic fact that these executives make $15,000 an hour compared to the $8 an hour of the custodial staff. Somehow this legal and even celebrated model of wealth distribution does not deserve our condemnation, but the latest scandal or public humiliation does. The great trauma of the economic meltdown of 2008, I believe, is only now descending upon us. That trauma has to do with the recognition that capitalism is still here, that despite the most significant economic crisis since the Great Depression the very structures in place since capitalism's inception have not substantially changed. It is brutal enough to lose one's job or one's home due to the crisis, but when very little changes in the process then we lose on both fronts. We lose our savings and our exhilaration, if not our joy, at watching the system give way—not to mention our desire, however unconscious, for the world to be organized differently.

This rethinking of crisis returns us to the anecdote about the psychoanalyst and to the problem of time. If crisis is always at work and not a sign that something has gone wrong, then how might we understand the *time* of crisis? Crisis used to be defined by its short-termness—requiring a decision on the spot, with no possibility of deferral, evasion, or repression. A crisis means we can, perhaps, suspend our usual rules and ethical standards because we must "act now!" But something has happened on the way to the shelter. The bombs have been launched, but they are suspended overhead, allowing us to continue on with our lives under the shadow of destruction. If crisis is always already with us, if it is the rule of the system rather than its exception, then in what cases could we justify suspending our principles? Would we have to live by them all the time? Or would this new temporality of crisis demand the rethinking of our society's cherished principles altogether?

Rethinking the meaning of crisis not only affects our temporal experience of the present, but also colonizes and preempts claims upon the future. Paul Virilio once argued that what defines the post–Cold War moment of accelerated communications and military technologies is a deadly simultaneity, a temporality in which a target is destroyed the moment it becomes visible. "When a missile threatening in 'real time' is picked up on a radar or video, the present as mediatised by the display console already contains the future of the missile's impending arrival at its target."[1] To see the enemy is to have already killed him. This logic intensifies an older paradigm in which one first scopes a target before trying to kill it. The converse logic holds that one kills the target even though one does not see it. Instead of precision bombs, we have under this converse logic carpet or cluster bombs and casualties, intended and unintended, that are identified after the fact. The atomic bombs dropped on Hiroshima and Nagasaki, as well as the improvised weapons detonated by suicide bombers, also fall into this lethal category.

This series of military logics, however, requires one more turn: today one kills the target but it does not die, at least not immediately. The target has been killed but has yet to die. There is a time warp at work, one in which the future has already come and is still to come, a double future. We can think about ecological predictions that forecast the end of natural resources and how it is already too late to reverse this categorical trend, or we can return to the military realm and consider how posttraumatic stress disorder attacks the soldier in the future for acts committed or witnessed in the present. In both cases there is a looping of time in which the future is spelled out in advance, granting to the meantime an impossible location that is heading somewhere and nowhere at once. But this is not a simple fortune telling, a determination already decided and thus eliminating contingency. This is, rather, a radically different experience and operation of time, one in which categories such as determination and contingency are refunctionalized. Or, even more to the point, this is an experience of time that enables the very shifting of how time works.

The paradigmatic condition illustrating the already dead is that of the medical patient who has been diagnosed with a terminal disease only to live through medical advances that then turn the terminal illness into a chronic one. The disease remains life threatening, still incurable, even though it is managed and controlled, perhaps indefinitely. The patient is now afforded a meantime that functions like a hole in time, an escape route to somewhere else *and* a trap door to where he began. Freud could only anticipate this double future in 1937 in a letter to Marie Bonaparte, written two years before his death and while he was in great physical and psychological pain (due to oral cancer and Nazi advances in Austria, respectively): "In order to find all of this bearable, one must remind oneself constantly that one really has no right to be living any longer."[2]

Freud's statement is symptomatic of his moment, while something qualitatively different is at work today. Of course, our present shares much in common with the modernity of Freud's time, but to tease out the differences we might want to invert Freud's statement: in order to find the current moment unbearable, we must remind ourselves that we really have the right to die. This is less about euthanasia, masochism, or legalized suicide than a certain relation to time, a relation to the present and the future, as well as the capacity to shape these temporal realities. Whereas Freud's project was to make the unbearable bearable by way of psychoanalytical practice, ours is to make the bearable unbearable by any means possible or, and perhaps more important, by any means impossible—unbearable to the extent that we cannot help but act toward another way of feeling and being, and impossible to the extent that we cannot help but act even in the face of an unavoidable fate.[3]

What is the relation between unbearability and revolution? Or between bearability and the status quo? Or between deadening environments (both social and ecological) and our relationship to death itself? This is where politics, culture, and medicine come together, especially around the concept of the already dead and the problem of time. We have entered a new chronic mode, a mode of time that cares little for terminality or acuteness, but more for an undying present that remains forever sick, without the danger of sudden death. The maintenance of the status quo becomes, if not quite our ultimate goal, what we will settle for, and even fight for. If the system cannot be reformed (the cancer eradicated, the ocean cleaned, the corruption expunged), then the new chronic mode insists on maintaining the system and perpetually managing its constitutive crises, rather than confronting even a hint of the terminal, the system's (the body's, the planet's, capitalism's) own death. In this work, I produce this concept of "the chronic," place it in the context of late capitalism, and track it in the political-economic, cultural, and medical realms. Departing from how changing medical practices (dominated by a new paradigm of targeted drug therapy and biotechnology) have reconfigured standard notions of "cure," as well as the meanings of "acute," "terminal," "crisis," and "meantime," I explore the radical and reactionary effects of this temporal shift.

Like cure in the medical realm, revolution has become the leper category for politics and culture. Revolution is thought to be old-fashioned, an embarrassing desire, hopelessly utopian, a mistaken objective in need of quick and certain displacement. The phrase "political revolution" sounds only slightly less ridiculous today than its ugly, disreputable cousin, "cultural revolution." The new chronic mode in medicine, in which the utopian desire to cure is displaced by the practical need to manage and stabilize, if not preempt the disease altogether (practiced in fields as varied as oncology, HIV, and psychiatry), is also at work in politics and culture. I am highly skeptical of this mode. At bottom, my argument is that although there is a progressive aspect of this current drive for management and preemption—a drive to transform the terminal into the chronic that is perhaps most obviously exemplified by the use of life-saving medications—there is also a reactionary dimension that effectively colonizes the future by naturalizing and eternalizing the brutal logic of the present. This baleful dimension is most effectively revealed when we analyze how the logic of "chronic time" works today in different cultural-political realms. If the possibility of death is removed, if the terminal cannot be even considered or risked, we effectively rule out certain courses of action in the present whose ends cannot be known in advance (precisely because we cannot know if they will end in death or the death of the present system). To remove the possibility of death and settle for the new chronic is to choose the known limits of the present over the unknown freedom of the future.

This leads me to the concept of the already dead as a means of rethinking the relationship between life and death today. I will argue two seemingly contradictory positions: first, that we remove the distinction between life and death, and second, that we simultaneously retain the relative autonomy of life and death. This removal allows us to engage the most pernicious ideologies of life and death. Their separation, in which death is figured as the great terrorist from beyond, requires and justifies the brutal, here-and-now sacrifice of our planet and species in the opportunistic name of rescuing our planet and species. "Better dead than red" was a Cold War way of exploiting this logic (in which communism was represented in the West as even more deadly than death itself and, therefore, requiring the sacrifice of all sorts of freedoms). More contemporary examples include the dreadfully high number of civilian casualties in the wars in Iraq and Afghanistan, wars conducted in the name of protecting humanity and civilization, and the gun-to-the-head corporate bailouts following the financial meltdown of 2008. Such political malfeasance is facilitated by, if not conditioned upon, a relationship to death in which death is radically separated from everyday life, and in which the existential terror of death is exploited by figuring it solely as dystopian fantasy.

Stressing the autonomy of death is essential to the consciousness capable of imagining radical change. Death is the pure form of radical change, and once our deaths are taken away from us in the name of the chronic then so is our capacity to imagine other radical possibilities, such as cure and revolution. Our right to die, perhaps not unlike Freud's right to live, is our right to dream—and live in—a radically different present than the one we now inhabit.

A growing number of people today are compelled to live in what I call the global abyss, a no-man's-land that opens up within global capitalism and forces our institutions and thinking to break down. What splits open this broken space are the convergences and contradictory imperatives of both the nation-state and the global system. Whereas the nation-state demands a certain economic outcome, the global system demands another. Whereas the nation-state demands a certain political affiliation, the global system another. Whereas the nation-state demands a certain cultural consciousness, the global system demands yet another. No doubt, there are key points of convergence in which the demands of both are in perfect coordination (it only takes a quick glance at the People's Republic of China to recognize that a strong nation-state can be in perfect coordination with the global system). And there can be no doubt the world is heading toward greater convergence and supplementarity. But right now we live at a moment when the logics of the nation-state and global system are crashing into each other, not to mention the crashing together of different nation-states and different components within nation-states, all occurring within the context of a transformed global space. These collisions leave in their chaotic wake an ideological formation that functions not unlike the new chronic.

The new chronic extends the present into the future, burying in the process the force of the terminal, making it seem as if the present will never end. Likewise, the global abyss extends all over the world, obscuring the place where globalization processes might end, making it seem as if this entangled national-global space (this functional crash of various uneven components) will never stop, making it seem as if we will be forever stuck in this meanspace. One cannot imagine a place where the processes of globalization stop. And by "stop" I do not mean those places where these processes have not reached, the so-called undeveloped or underdeveloped nations, but a beyond to globalization where a country like the United States is no longer global. It is as impossible to imagine a post-global United States as it is to imagine a future that is not a chronic extension of the present. If this new radical temporality or spatiality were imagined, it would be done so only on the order of a grand apocalyptic finish—a nightmarish fantasy that functions to limit the imagination. This book explores this new ideological configuration in various cultural-political spaces.

I conclude my analysis by returning to the already dead. It should come as no surprise that questions of life and death have been placed at the forefront of the contemporary intellectual agenda. Culture, politics, and medicine have shifted over the past few decades along with the reconfigurations of global capitalism. With these shifts have come four of the richest concepts of intellectual culture. These are Slavoj Žižek's concept of the "undead," Jean-Luc Nancy's "living dead," Giorgio Agamben's "bare life," and Margaret Lock's "twice dead." And each of these concepts moves laterally to overlap with a series of concepts as varied as Lacanian drive, Derridean *différance*, Marxian revolution, Foucauldian biopower. To differentiate the already dead from these other concepts, I will begin by reading them in terms of the problem I have already established: how to extinguish the division between life and death while at the same time retain the autonomy of the two states. In one form or another, these four concepts and that of the already dead all attempt to resolve this problem by how they theorize death. But death is always and ultimately a problem of time; all theorizations of death are at once theorizations of time. I produce the already dead, therefore, in order to foreground this relation between time and death and to explore the various political possibilities that emerge when we think of them together.

The new chronic and the global abyss are ideological formations special to contemporary capitalism. These formations wholly structure our relation to reality; they build their assumptions into the way bodies and the environment are managed through the logic of capital. They shape our reality to such an extent that they shape the very categories of time and space. The already dead, however, resists full inscription into these ideological formations and shakes up the possibility for an active political resistance. The already dead is not more real (or less ideological) than that of the new chronic or the global abyss, rather it erupts from inside these ideological formations to change everything. The concept of the already dead suggests a future beyond the temporal constraints of the new chronic and the spatial constraints of the global abyss—a future content beyond its own form. The already dead, therefore, do not constitute a political movement in the traditional sense. Rather, they portend a political consciousness that can inspire and inform political movements. The already dead already inhabits revolution—that is, they inhabit a revolutionary consciousness informed by a certain way of living in time and space, and in relation to an unknown and unrealized future.

The new chronic, the global abyss, and the already dead are concepts inspired by my own experience. I was diagnosed with a certain form of leukemia, which at the time of my diagnosis was understood to be terminal but is now considered chronic by way of a new targeted drug that promises to manage the illness far into the future. Of course (of course!), I am quite relieved about this timely turn of events and my returned future. But I refuse to temper my critique due to such a lucky reversal. The drug I take to manage my leukemia binds itself to certain chromosomes, thus interrupting the mutation that would otherwise lead to my certain death. This drug costs such an absurd amount of money (over $45,000 a year) that the expense caused my application for Canadian permanent residency to be initially rejected. But the drug also performs a crucial ideological function that shapes the way we come to terms with everything from politics and culture to the most banal aspects of our everyday lives. This gets to the heart of my study: the paradigm shift in which the older notion of cure has been replaced by a newer notion of "management" (the supplanting of the more radical dimensions of both cure and death by the dimension of the new chronic). This shift has far-reaching and crucial implications—and not only for medicine. How we understand and believe in cure shapes how we understand and believe in a whole host of nonmedical problems, just as the way we understand and believe in revolution shapes how we understand and believe in nonpolitical problems. The very way we imagine the future and act in the meantime is at stake.

I will not invoke my personal diagnosis and the ensuing immigration battle with the Canadian government until I discuss the global abyss—all the while acutely aware that whenever this space is jumped (from the local to the global, from the individual body to the social one, from the personal to the impersonal, from a singular event to the larger series of events and back again) something gets forced, something gets lost, only to reappear again, in-between everything else. Whatever this something in-between is, it is as supple as hope and as hard-felt as a needle in the arm. But this thing is also another way of describing the new chronic, the global abyss, and the already dead. For although these three conditions do not capture everyone, although they are, for many, a safe distance away in a refugee camp, in a makeshift palliative care tent on the edges of a migrant town, or take the form of an unlikely medical diagnosis, these concepts and conditions come together to represent the paradigmatic space of our present situation—one that envelops us all. Though we fall at different speeds and with differing degrees of recognition, we are freefalling together in the global abyss and living together as the already dead. Where we will land, or if we will ever land, is hard to know, but what is certain is that along the way something will be broken (indeed, something has already been broken) and be revealed as so many possibilities.

This takes us back to the opening anecdote about psychoanalysis and the reinterpretation of crisis. What happens when we push the anecdote to its logical end? Or to its logic of the end? When considering whether to end the analytic sessions, the patient asks the analyst if the present time is right. "Well, how are you feeling?" asks the analyst. "Awful. My life is in a shambles and I have nowhere to turn. I'm in the middle of a real crisis." "In that case," says the analyst, " you'd better stop right away." Or does it go like this: "I feel great. My life is wonderful and my work satisfying. Best of all I understand myself and feel more awake." "Oh in that case," says the analyst, "we better continue." There is something missing in both punch lines, or at least there is something insufficient. Is there a way to hold—at the same time—both possibilities? Is there a way to both continue and break off the analysis? This is what the concept of the already dead wishes to do—to make death continuous with life, while maintaining the break of death. Again, this is not restricted to issues of living and dying, but provides a more general mode of engagement, one that I will mobilize to negotiate the various problems that emerge throughout this work.

How to challenge and resuscitate the concept of cure, while not giving oneself over to management? How to challenge the new chronic by preserving the intensity, and even the utopian quality, of the terminal? How to retheorize revolution today, while not falling into liberal reformism? How to question contemporary theorizations of culture while not fetishizing practice? And, finally, how to question the celebration of globalization while not embracing the nation or nationalism? It is the logic of the already dead that suggests the possibility of squaring these circles. This logic loosens and even undoes (however impossibly) the binaries, while tightening and reinforcing the political significance of their individual terms. We can have it both ways, if only in the meantime. And the meantime might never end. Or, and if we are even luckier, the meantime will end, bringing with it something else that, at least right now, is beyond our capacity to imagine, or to desire. But—and this is when the circle seems to close—this desirable end of the meantime (this revolution) can open only if we do not directly imagine or desire it. Instead of letting the complexities of this philosophical circle discourage us and put an end to our dreams for a better world, the concept of the already dead flashes the possibility of how we might keep it from closing, thus keeping alive this desirable but not to be desired utopia.

PART 1

The New Chronic

There are moments in history when time seems to stand still, when the layers and cycles and rhythms snap back into a single instant, stinging us in a way that is not quite painful but undeniably felt, like a rubber band whipping back against our skin. If you can imagine this, then try to imagine its opposite—the stretching of time, the extending, flattening, and rolling thin of time, the dull soreness of a meantime with no end; and if you can imagine this (if not feel it), then you have come very close to understanding how the dominant mode of time works today.

We have entered something like a new chronic mode, a mode of time that cares little for terminality or acuteness. Every level of society is stabilized on an antiretroviral cocktail. Every person is safe, like a diabetic on insulin. A solid remission, yes, but always with the droning threat of relapse—of collapse, if not catastrophe, echoing back to us from a far-off future or from the memory of a distant past. So monotonous and stripped of urgency is this mode of experience that there is little reason to agonize, as long as the pain is managed and the possibility of any future change is repressed. All signs must remain "unremarkable," as the doctors like to say when the health tests come back negative. These medical metaphors (fairy tales, really) are not accidental, for it is medicine that leads this new mode of time, a mode in which we relate to death like a child to his future—with dreamlike indifference and from a safe distance. But, as we all know (and certainly for those who, in one of those untimely instances, lose their lives in earthquakes, in bombings, and in health emergencies in villages and cities around the globe), the future cannot be put off, crisis and disaster cannot be totally managed, life can never be safe, and we do not all experience time—and certainly not the political effects of time—in the same way.

My point in raising the issue of time and the medical is not to pursue the raw and heartbreaking question of unequal access to, and quality of, health care the world over or to talk about the social causes and uneven effects of disaster. My purpose is to claim that time itself constitutes a privileged path to understanding our present moment. Not just time, but the dominance of a certain mode of time—one that resists viewing the future as anything other than an extension of the present and one that can be tracked in various levels of the present global formation.

Changing medical practices have shifted standard notions of the chronic and the terminal, thus shifting how we manage the present and the future and, indeed, how we think about and *feel* time. Likewise, within political and cultural theory (and practice) a targeted approach to the present, a certain prescriptive mode that limits itself to targeting individual symptoms, is becoming dominant. Site-specific and occupied by single issues, this "immanent" mode celebrates the guerrilla fighter or performance artist who seems to disappear dreamlike at the very moment of his or her radical act—all in real time and all without an apparent claim to something larger.

The chronic, therefore, is not solely a medical category. It is as much about the psychological and emotional relation to illness as it is about life-saving medications, as much about economic accumulation and geopolitical relations as it is about the latest definitions of death. It is as much about the possibilities of aesthetic representation as it is about the latest medical-imaging technologies.

But this widening of the chronic does not only lead to a political-economic, psychoanalytic, or cultural analysis of medicine. It leads to analogies and shared formal logics among the most disparate aspects of our lived experience. The elaboration of these analogies (the elaboration of the problem of analogy itself) is central to my argument, for there is an overarching logic of time that jumps from one level to another (medical, cultural, political) and, when highlighted, reveals key insights. But overarching does not mean transhistorical or universal; rather, the current logic of time is specific to the present moment of global capitalism. Indeed, capitalism always exceeds itself; it is not only about capital (not only, that is, about production and exchange, value, labor, class, and consumerism). Capitalism is both the inside and outside of our being, informing even its (capitalism's) own undoings, its resistances, its alternatives—its unthought and unacted. Capitalism digs its own grave, yes, but not just for the moment of its death; it does so also as a productive act in the present, an act that can, perhaps, teach us something about how life and death, as well as any number of other apparently opposing terms (health and illness, for example), are not always or necessarily chronologically ordered, one term preceding the other like so many days on a calendar.

The idea and desire we call "cure," and the idea and desire we call "revolution" are inextricably tied to each other, and today both of these categories are in freefall. A productive freefall, no doubt, and one that opens up new progressive possibilities, from life-saving medications to new social movements to the most ingenious aesthetic interventions. But there are crucial limitations as well, limitations that turn on a reluctance to engage the central question of radical change and what this reluctance means for imagining alternative futures and, more important, acting in the present. And this returns us to time, for it is through an attentive engagement with time that we can (however blindingly) see the stakes of our current moment, the medical stakes of the political, the political stakes of the medical, and the cultural stakes of the biopolitical. But we also enter into the singular realm of the human psyche, though perhaps from the back door, and must remember that none of this is elegant, none of these realms fit flush with the other. There is always something missing, always something that cannot be squared. As Chris Marker reminds us in the film *Sans Soleil*, "Who says time heals all wounds? Time heals everything but wounds."

To hear time as a keynote today (we have just established it is not a panacea) seems to force us down a philosophical rabbit hole, returning us to ancient brainteasers on the meaning of time or to the various clock technologies, from sundials to obelisks, from the seventeen-year libidinal cycle of the cicada to the beating of our own hearts. And then there is the modern history of theoretical physics, with its tenses and loops, its Möbius twists and infinite branchings all the way to the very end of time in superstring theory and canonical quantum gravity. But this focus on time (as a cutting-edge scientific problem or as one of the key philosophical puzzles persisting at least since St. Augustine) should not divert us from recognizing that something is going on today with time that marks a qualitative shift in how it operates on us, a shift that relates to new and old inquiries into time, surely, but a shift that cannot be reduced to science and philosophy. As already proposed, I name this shift, this dominant operation and experience of time, the new chronic.

Before beginning, however, there is one more thought experiment that I want to suggest: try to imagine what comes after globalization. If you find this difficult, if not impossible, then perhaps it is because imagining what is beyond globalization is like imagining what comes before or after time—a mind-bending exercise, indeed. But this inability to think beyond globalization is precisely one of globalization's most crucial ideologies. If one were to imagine what comes after globalization, then globalization would cease to exist, for globalization is precisely that category that incorporates and absorbs everything into its realm, especially the very thought of its own end.

One of the more interesting facts about the globalization debate is that almost all historiographical concern has been directed backward, so that locating the origins of globalization becomes the fundamental point of confrontation. Did it begin with the oil crises of the 1970s or the moment of political-economic integration after the Second World War? Should nineteenth-century colonialism or the expansion of fourteenth-century trade routes mark its beginning? When someone enters into the globalization debate they invariably smuggle into their argument an argument about origins. Every argument about globalization, therefore, is a theory of history. But such a theorization is almost always unconscious. Most thinkers directly engage the question of when globalization began or whether or not it has even started, but they rarely considered what opens up or closes down when history is punctuated in such a way. The question of "an end" to globalization is not to reproduce from the other direction this repression, but to reveal how the problems of history and time are at stake, how globalization, like certain manageable conditions, can admit an origin but not a termination point. To name an endpoint to globalization is to simultaneously do away with globalization as we currently understand it. The very category of globalization depends upon the dominance of a certain notion of time, the new chronic.

The new chronic, however, is not an entirely smooth and sealed temporality. It contains a fundamental crack within its formation, a gap that comes between its ideological work and the possibilities of human existence. Here I am reminded of the joke told by Slavoj Žižek about two security officers patrolling a city street after a military coup in Poland. The officers have orders to shoot and kill anyone out on the street after 10:00 PM. It is ten minutes to ten and one of the guards sees a man hurrying along and shoots him dead. The other officer, perplexed and worried, turns to his partner and asks why he shot too soon. "I knew the fellow—he lived far from here and in any case would not be able to reach his home in ten minutes, so to simplify matters, I shot him now."[1] The new chronic experience of time submits to this very logic, a logic that assumes that everything will remain the same as the present turns into the future. The reality that structures this "now" in time will be the same reality that structures subsequent "nows" (not only does this invoke a classic Aristotelian mode of how time works, but this mode continues to be pressed into political service today, an ideological exercise regime for reducing the imagination). No doubt, and as I will stress throughout my analysis, this reality is provoked by (and provokes) very real fears and vulnerabilities—an existential mode that privileges management over change and holds fast to rigid continuities while walking with only the most tentative and straightest of steps.

Notwithstanding the many progressive aspects of the chronic in the medical, political, and cultural realms, there are many reasons why such a mode is also debilitating and inadequate. In the context of our joke, the curfew could be lifted during the ten minutes before it is ten o'clock, transforming bureaucratically sanctioned efficiency into murder. There could be another coup. The guard could be fired in the meantime. The man could have moved closer to the city center. The point is that when contingency is removed from the present and the unknown (including the potential termination of the present) is eliminated from the future, we have devolved into a military state, one that in order to reproduce itself must shrink the imagination and squeeze dry the experience of time. The following is an exploration of the extent to which this military state is an epistemological and existential state of being (not necessarily an authoritative political structure) and inextricably tied to a dominant mode of time.

The Medical

Today, there is an emerging global culture of cancer in which very few doctors speak of a cure, and in which even the category of remission is starting to lose its value. Rather, the watchwords are now management and pre-emption. Instead of depending on the total removal of cancer by either cutting, burning, or chemically killing it (procedures that always leave the possibility of relapse), doctors regularly turn to drugs produced to manage cell growth and other procedures (such as stem-cell therapy) designed to pre-empt the very manifestation of cancer itself. In the trenches where medical researchers and clinicians—and pharmaceutical executives—work, cancer is quickly transforming from something to be cured to something to get along with, to manage with technologically advanced drugs that keep things in check. The war metaphors still exist, but now instead of carpet bombs and nuclear blasts, we have smart bombs and reconnaissance drones.

At the heart of such transformations is the radical expansion of detection technologies. For example, recent experiments with highly sensitive medical imaging systems revealed that a preposterously high number of people in the general population are walking around with certain forms of cancer. Whether or not the diseases will accelerate before their hosts die of something else is difficult to determine. This has generated a certain crisis of meaning for the radiologists employed to read medical images. The older paradigm under which many radiologists were trained to read (with specific interpretive strategies and narrative assumptions) no longer seems to work; it no longer seems to apply to the foreign texts the new technologies are now producing. This is also connected to what has recently been called overdiagnosis.

In the summer of 2008, the U.S. Preventive Services Task Force recommended that doctors stop screening men ages seventy-five and older for prostate cancer. Since these men, on average, will not live longer than ten years, and since prostate cancer progresses slowly, the study determined that more harm than good could come from screening prostate-specific antigen levels. The study effectively argues that prostate cancer is overdiagnosed in 29 to 44 percent of cases. Overdiagnosis occurs when a medical test picks up an illness that will not cause symptoms during the patient's lifetime. The study goes on to argue: "Because patients with 'pseudo-disease' receive no benefit from, and may be harmed by, prostate cancer screening and treatment, prostate cancer detection in this population constitutes an important burden." In this context, "pseudo-disease" does not refer to false positives (when detection technologies incorrectly indicate disease characteristics that do not actually exist), but rather to a disease that is correctly diagnosed, but clinically insignificant.[2] As might be expected, the responses to this study were severe and hostile, most coming from healthy men in their seventies and eighties who were offended by the statistical argument. "Isn't the whole purpose of futuristic medical care (including new testing and treatment breakthroughs) about beating—and changing—the odds," those in this demographic seemed to be saying. When cancer might be in all of us and only some of us are unlucky enough to suffer from it (or when cancer and other terminal illnesses are made chronic or are overdiagnosed), then the very categories of health and illness, benign and malignant, cure and relapse, and perhaps even present and future, become permanently confused.

The consequences of such confusion are especially acute if you are an insurance actuary, a health care bureaucrat, or even a future employee of a company relying on medical imaging. In the United States, for example, there is an emerging trend in which insurance companies and employers subject clients and potential employees to medical screenings, the interpretation of which determines their future, a determination based simply on the likelihood of illness, not even on the actual presence of illness.[3] But without a clear-cut category of preexisting illness or of illness itself, how does one determine risk? How does one plan for the future? And on top of this, the usual ways of determining the durability of a therapy through long-term results is radically changed when management (with its rapid modifications and unique-to-patient configurations) is emphasized over a cure. As usual, there is also a certain economics at work in which an emphasis on management serves certain financial interests, usually the interests of global pharmaceutical corporations whose therapies usually require long-term, continual use. An emphasis on prior notions of cure serve other interests, from the interests of those with a stake in traditional medical rituals to those invested in nationally based protocols of medical training and practice. And patients often get caught in the middle.

Of course, an oral therapy that targets certain genetic mutations, that turns off switches to control the overproduction of cells, is a gift for many who would otherwise die or undergo dangerous and painful invasive surgeries. Easier to mass-produce, these therapies also offer the potential for wide distribution in ways that in-hospital procedures do not. The problem here is that within global capitalism such drugs must be commodified and, therefore, inaccessible on an equal basis.[4] And this is not because there are too many greedy pharmaceutical executives. If these drugs were properly distributed (accessible to all), we would, quite simply, be in another economic system (this is less about whether a certain nation has a public or private health care system and more about a global medical industry that has no choice but to swear allegiance to the flag of commodity culture and relentlessly pursue intellectual property rights). The inequality of health care, especially when dominated by the pharmacological-therapeutic paradigm I am outlining here, is not what occurs when capitalism goes wrong, but when it goes right. In any event, cure, as a condition separate from disease, is quickly being removed from the medical vocabulary. This new chronic condition is both an effect and cause of the new dominance of prescriptive medicine.

COMPLEX CHRONIC DISEASE, PRESCRIPTION, PLACEBO

The emphasis on the chronic has produced a new disease called Complex Chronic Disease (CCD). Bridgeport Hospital in Toronto has recently re-branded itself as the only hospital in Canada to specialize in treating CCD. "Our hospitals have become very good at saving lives," Bridgeport advertises. "Thanks to modern medicine more and more people are surviving serious conditions such as cancer, heart disease and stroke. This creates the great healthcare irony of the 21st century: we're living longer, but living with these lifelong or chronic diseases. What's worse, most Canadians will be living with not one, but several chronic conditions." The copy of this particular advertisement ends with the following one-liner: "Our goal is to give you back your life after modern medicine has saved it."[5]

Seventy percent of Canadians over the age of forty-five, more than 16 million people, are living with two or more chronic diseases. The Cana-

In a recent rebranding campaign, Bridgeport Hospital in Toronto introduced a new disease called Complex Chronic Disease (CCD); they claim to be the only hospital in Canada that specializes in treating CCD.

dian government spends over $150 billion a year on health care, and as much as 70 percent of that is directly related to chronic disease.[6] With over 60 percent of lost productivity attributed to CCD, these new hybrid diseases (which Bridgeport glibly names "osteocanceritis" and "neurodiabesity") are steering health care toward a focus on prevention and management.[7]

One key aspect of CCD care is to negotiate the way different pharmaceuticals interact with each other, since until recently no one considered whether, say, AZT (a drug for HIV) was counterindicative with Gleevec (a drug for chronic myelogenous leukemia). No one considered this because no one had to consider this: one was either going to die of AIDS or of leukemia, but not live with both. Prescriptive medicine that turns formerly terminal illnesses into chronic ones requires a sustained, long-term drug therapy and sustained, long-term compliance by the patient. Indeed, disease management based on medical prescriptions requires a careful reconsideration of how the prescription itself functions under these new circumstances.

A medical prescription is written in advance by a doctor as a way to treat a specific condition in the present. It is a written directive that responds to specific symptoms and whose main objective is to keep the symptom from accelerating into full-blown illness. In general, it is not programmatic, neither is it geared toward eradicating the cause of illness. Prescriptive medicine begins with the axiom of health management: as long as illness is managed in the meantime, we do not have to worry about curing for the future. The prescriptive meantime becomes the permanent destination rather than a temporary moment of development. Control over the symptom is the gold standard, and even though the underlying illness is still understood as malignant, the management of the malignancy is now possible in new ways. What all of this means, for example, in relation to traditional Chinese medicine and its focus on causes instead of effects, wholes instead of parts, is significant and illustrates how global medical trends and discourses crash into national and local ones, not to mention how complementary and alternative medicine itself is becoming commodified and submitted to the same logic as so many cutting-edge biopharmaceuticals.

This prescriptive moment emerged during the 1980s and hit its stride at the end of the century, with the successful sequencing of the human genome. This might sound counterintuitive, given the long history of prescriptive medicine, beginning as far back as Babylon, as evidenced by clay tablets on which prescriptions and directions for compounding were recorded. Still, to use the phrase "prescriptive medicine" today is to invoke the qualitatively different role of the prescription within global capitalism, that is, the prescription as a phenomenally profitable commodity of one of the most successful capitalist industries the world over. And, yet, contemporary prescriptive medicine is not only different from premodern medical prescriptions, but is different from the pre– and post–Second World War moment of the pharmaceutical industry's great expansion. From the discoveries of penicillin and insulin in the 1920s to the release of Milltown and other breakthrough tranquilizers in the 1950s, this growth was generated though a sort of scientific luck. "The origins of virtually every class of drug discovered between the 1930s and the 1980s can be traced to some fortuitous, serendipitous or accidental observation," writes Dr. James Le Fanu in his book *The Rise and Fall of Modern Medicine*.[8] With new knowledge provided by the Human Genome Project about how cells grow and interact with each other, however, researchers are now able to directly target certain mutations by blocking specific pathways and turning off the signals that cause the overproduction of cells—all the while leaving healthy cells alone. Regardless of how overhyped and exaggerated scientific and financial utopian claims of biotechnology are (in fact, most biotechnological startups fail and less than 1 percent of first-tier biotechnology companies in the United States have become profitable), the new paradigm of management is rarely put in question.[9]

Following the famous Bayh-Dole Act of 1980 in the United States, business partnered with academia to grant academic scientists an automatic right to patent and sell their federally funded work to private corporations, effectively allowing these new alliances to steal basic science for profit. The antiretroviral triple cocktail (which has been so successful in managing HIV) is just one example of how taxpayers the world over paid for the basic science, while transnational pharmaceutical corporations reaped enormous and easy profit. "In 2001 the ten US drug companies in the Fortune 500 list ranked above all other US industries in average net return, whether as a percentage of sales (18.5%), of assets (16.3%), or of shareholder equity (33.2%)."[10] Judy Law goes onto explain that, "In 2002 the combined profits for the ten drug companies in the Fortune 500 ($35.9 billion) were more than the profits for all the other 490 listed businesses put together ($33.7 billion)."[11] While since 2003 the oil industry and defense contractors have led the way in profitability (a change unsurprisingly coordinated with the beginning of the Iraq War), pharmaceuticals (along with oil and weapons) is still one of the great success stories of capitalist development.

Most critics go after big pharma by focusing on the "me-too drugs," drugs that are similar to those already known, with only minor differences so as to circumvent patent laws. These critics also tend to focus on the huge marketing and promoting budgets of pharmaceutical corporations (budgets almost twice as large as those for research and development), the unequal access to prescription drugs that is most evident throughout the global south, the corrupt suppression of scientific data and research that contradicts intended claims (for example, the Vioxx scandal), the aggressive lobbying of doctors and health care bureaucrats and the fear induced in vulnerable consumers, and the fact that as pharmacotherapy becomes dominant it reconfigures medical practices in ways that marginalize not just other therapies and the development of other therapies but various social solutions to illness, such as alternative ecological and labor practices.[12]

Many on the political left have switched their positions over the years. The critique of mental health drugs, for example, began as a deep skepticism of antidepressants. Critics argued that depression was being discursively produced and exaggerated so that pharmaceutical corporations could commodify the treatment and turn huge profits. One must reject the category of depression, so the critique went, and wean oneself off the drugs that so effectively reproduce the capitalist system. This first critique responded to the early euphoria (at the beginning of the 1990s) around selective serotonin reuptake inhibitors, such as Prozac, and key changes to the fourth edition of the *Diagnostic and Statistical Manual of Mental Disorders* (DSM) in 1994. By the end of the decade, a different critique could be heard: depression was not an opportunistically produced disease by insidious pharmaceutical corporations and an all too cooperative medical establishment, but a very real illness that was a direct result of late capitalism's alienating effects. How could we not be depressed when money is the measure of all value and our most private desires (sexuality) and our most public resources (nature) are submitted to the logic of capital like so many other goods and services? The circle to be squared was now how to respect the horror of depression and even the successful pharmaceutical treatments of it, while at the same time retaining a trenchant critique against the commodification of medicine. Could Prozac exist without Eli Lilly?

One response to this question is to appeal to the powerful effect of placebos. Various studies of depression, for example, reveal that upward of 50 percent of patients get better when given placebos and that the rate of relief from depression generated by the placebos is not significantly different than the rate of relief generated by a whole range of antidepressants.[13] In his book *Anatomy of an Illness as Perceived by the Patient*, Norman

Cousins relates an experiment in which both arms of a study of patients with bleeding ulcers were given placebos.[14] Those in the first group were told that they were receiving a highly successful drug that worked in almost all cases, while those in the second group were told that they were receiving an experimental drug with very little data. Seventy percent of those in the first group received significant relief, compared to only 25 percent of those in the second group. For many, such studies undercut the legitimacy of pharmaceuticals and even provide a model for noncommodified, noncorporate drug therapies that work, namely cheap and copious sugar pills.

But we could just as well understand such studies as confirming the power and success of the pharmaceutical model. Placebos only exist and work in relation to the actual drugs themselves, to the larger system of prescriptive medicine with its packaging and costs, and to discourses of trust and efficacy (not to mention hope and desire) placed in the model. Also, one must not know one is taking a placebo for the effect to work. Moreover, the doctor who prescribes the placebo must not know—to be most effective the act must be what is called "double blind."

In 2008, a placebo for children was marketed in the United States in the form of chewable, cherry-flavored sugar pills called Obecalp ("placebo" spelled backward). The ethics and efficacy of the drug were immediately called into question by various experts, from child psychologists who contended that parenting based on deception is not healthy to pediatricians claiming that placebos are unpredictable. Medical ethicists wondered why anyone would want to produce early habits of pill popping in children, and clinical scientists believed that the desired effect of the placebo would be compromised if the parent knew she was giving her child a placebo.[15] In response to these criticisms, Jennifer Buettner (the person who came up with the idea for the pill) argued that in fact the placebo could help wean children off drugs, since it is nothing more than a sugar pill. "The overprescription of drugs is a serious problem, and I think there needs to be an alternative," she reasoned.[16] If one takes a placebo believing that it is the "real" drug, how does that offer an alternative model? And this leads to a whole set of related questions, such as, can a placebo have the same side effects as the actual drug? And what about the "nocebo effect," when a patient's lack of confidence in a drug weakens the response? "I will harm," the translation of the Latin "nocebo," does not have the "pleasing" effect placebo's etymology promises.

If the placebo only confirms the legitimacy of the larger pharmaceutical model, however much individual studies delegitimize the exceptional claims of particular drugs, then we might want to ask another question that returns us to the chronic and the role of cure: What is the time of the placebo? And how might this time, and the time of the pill in general, contrast to the time of other health care models? These questions open up the psychological dimension and the differences between the pharmaceutical management of mental illness and the so-called talking cure of psychoanalysis.

In a book titled *Talking Cures and Placebo Effects*, David Jopling argues that psychoanalysis and other forms of psychodynamic psychotherapies are themselves placebos, meaning that they may in fact successfully heal patients, but they do so not because of any techniques or methods particular to their treatments (from special modes of interpretation to transference). Rather, the talking cures work by way of placebo effects that rally the mind's native healing powers. What disturbs Jopling the most about talking cures are the claims they make to "a unique set of characteristic ingredients" that are nothing more than pseudo-insights, explanatory fictions, and self-deception—a wholesale disregard for the truth of what ails an individual. The appeal to pseudo-scientific explanations is unethical and dangerous, according to Jopling. "What looks like *bona fide* insight, or self-knowledge, or a genuine realization or a new and more empowering way of looking at oneself, may in fact be ethically calamitous."[17]

What concerns Jopling is not the "truth" of various mental health conditions, but only that psychodynamic psychotherapies have little regard for that truth and make philosophically and medically spurious claims. If the talking cures work because of placebo effects, then they deserve criticism based on how expensive they are to individuals and public budgets, how expensive and time-consuming psychotherapy training programs are, and how unethical they are as they deceive the public "with false, exaggerated, or unsupported claims about therapeutic effectiveness and explanatory success, and how they don't properly treat very serious mental illnesses."[18]

Jopling's work fits squarely into the genre of antipsychoanalysis and Freud-bashing that has a long history from Karl Kraus to Karl Popper to Frederick Crews. But what is peculiar is Jopling's insistence that the talking cure is flourishing at present, that it is "never more popular, more available, and more in demand, than it is today."[19] In terms of training programs, one cannot compare the amount of funding for pharmaceutical programs to the funding to train those who practice any of the types of therapy Jopling criticizes. Moreover, within the larger field of psychiatry, psychoanalysts are clearly on the margins, compared to neurobiologists and psychopharmacologists. One study revealed that the percentage of visits involving psychotherapy declined from 44.4 percent in 1996–97 to 28.9 percent in 2004–2005.[20] My interest is less in these statistics and more in what such a dramatic shift suggests about larger ideological shifts within mainstream culture.

PSYCHOANALYSIS, CURE, PRAXIS

The reversal of fortunes over the past thirty years that psychoanalysis has experienced, once the preeminent "talking cure" (whether a cure for psychological symptoms or a cure for the desire for a cure itself) and then the embarrassing uncle that the field of psychiatry would rather keep hidden, does relate to the decentering of the concept of cure today and to the tremendous growth of psychotropic medications. Many are familiar with Freud's quip about cures in *Studies in Hysteria* (1895), written with Joseph Breuer: "Transforming hysterical misery into common unhappiness." But it is the next line that is often forgotten. "With a mental life that has been restored to health, you will be better armed against that unhappiness."[21] Indeed, common unhappiness is not the end (for this would be *management*, this would coincide with the temporality of prescription), but neither is it the good-enough cure. Rather, uncommon unhappiness is a means to an end, a means to a more productive struggle with unhappiness and even, possibly and however momentarily, to happiness itself. For Lacan the desire for a cure was limited, but crucial nonetheless. "All the same, cure always seems to be a happy side-effect . . . but the aim of analysis is not cure."[22] What I wish to tease out here is a psychoanalytical theory of cure, but one that exceeds the curing of neurotic symptoms, to reach the wider philosophical and political stakes of the category. Without cure charging desire, charging the desire of the patient and the analyst, charging the desire of psychoanalysis itself, we have left the realm of psychoanalysis. Perhaps this is another way to distinguish between psychoanalysis and psychotherapy: analysis is no longer possible without some mobilization of cure (even as a negative category). Therapy, in contrast, can manage quite successfully without that mobilization. Indeed, psychotherapy is at its best when the problem of cure is not at issue. Still, the most common distinction between analysis and therapy is understood in terms of how time is organized, from the frequency of sessions to the duration of the treatment, from the time-consuming act of free-association to the very way the past, the future, and the "timelessness" of the unconscious are invoked.

For Freud, time was inextricably tied to cure. In a 1910 essay, "On Wild Psychoanalysis," Freud refers to an inept psychoanalysis conducted by a young doctor who does not appreciate the role of time.[23] A recently divorced woman in her late forties and suffering from anxiety consulted Freud. The woman, skeptical about the young doctor's diagnosis, wanted Freud's opinion. The young doctor, himself enamored of psychoanalysis, had explained to the woman that her problem was caused by sexual deprivation and that she should choose from among three courses of action: return to her husband, take a lover, or masturbate. Freud was critical of the young doctor for many reasons (for starters, Freud contended that the young doctor had an insufficient understanding of what the sexual meant in psychoanalysis). But what most disturbed Freud was that regardless of whether the proposed cure was correct, the form by which it was delivered was all wrong.

Psychoanalysis is about the time it takes to trigger resistances and activate emotional attachments (most specifically the transferential relationship between patient and analyst). Even if the analyst were to correctly diagnose a patient during the first session, the patient would almost certainly be unable to do anything with the information. The point is not that the woman received bad advice, but that the timing of the advice was bad. Or, more significantly, the *time* of the advice was bad. The young doctor's advice assumed a certain logic of time that understood a linear trajectory from the moment of ignorance to the moment of knowledge to the moment of changed behavior. This assumed progression is precisely what psychoanalysis calls into question.

The revelation of a blind spot does not ensure progressive change. The unconscious is not something that can be exposed like an X-ray. Such an interpretive strategy—reading the X-ray—responds to a different narrative logic than that of psychoanalysis. This search for the buried treasure evinces certain theoretical assumptions about representation that allow that search to be assimilated by the new chronic in a way psychoanalysis resists. Such a mechanical act of exposure, as Freud reminds us, is like distributing menus to the hungry during a famine.[24] But here a distinction must be made between a specific unconscious desire and the unconscious itself, between specific blind spots and the blind spot itself. On revelation, Freud writes, "This is an outdated idea, based on superficial appearances, that a patient's suffering results from a kind of ignorance, and that if only this ignorance could be overcome by effective communication . . . , a recovery must follow." And Freud goes on, "But the illness is not located in this ignorance itself, but in the foundation of ignorance, the *inner resistances* that are the cause of the ignorance and continue to sustain it."[25]

Are we prepared to take the cure? The whole point is that we are not. The old joke about how many analysts it takes to screw in a light bulb (one, but the light bulb must really want to change) is misleading.[26] Perhaps the more appropriate answer is: "Time's up, we'll continue this tomorrow." This Lacanian punch line is a play on how the "variable-length session" (psychoanalytic sessions that can last from a few minutes to a few hours) frustrates the dominant form of desire and demonstrates the understanding that change is not possible without engaging our dominant habits of relating to time and the way we "really want"—not the way we really want this or that thing or person, but the way we want, in general. The punch line is double: it works on the level of content (within the genre of light bulb jokes), and on the level of form. It engages the joke itself, how the system of jokes work, how our expectations work. The punch line is that there is no punch line, and this gets to the foundations of the joke, to comedy itself, which, at very bottom, is bottomless, a void, a "nothing," save our own desires to express so much hope and despair in the form of laughter, the more hysterical the better. But even this descent into the inner workings of the form is not enough, for if it were "then all you would need for a cure would be for the sufferer to listen to lectures or read books."[27] This is where the praxis of analysis comes in.

Two key struggles mark Freud's work. First, how to instantiate psychoanalysis as a system, with all of the requisite protocols and rules and need for legitimacy, while at the same time not betraying the contingent nature of psychoanalytic treatment, the fact that psychoanalytic practice must always break with any prevailing orthodoxy. And second, how is it possible for Freud himself to be analyzed (the only path to being an effective psychoanalyst) before there are any trained analysts—let alone before there is that thing called psychoanalysis? How, in other words, can Freud be anything but a fraud and psychoanalysis be anything but fraudulent, given that its father broke the first and most fundamental rule—that every analyst should be formally analyzed? Indeed both of these circles could only be squared by focusing on their very impossibility and privileging the antinomies themselves, that is, by privileging the tension between theory and practice, the vital irresolvable problem that drives the coming into being of psychoanalysis.

As for rules and protocols, we can turn to "Analysis Terminable and Interminable," one of Freud's later essays about whether it is possible to shorten analytic treatment. From the title alone, we can identify an unexpected argument. How can something be, at one and the same time, both terminable (coming to an end) and interminable (never ending)? And for our own purposes, we might want to develop the similarities and differences between what Freud is calling here the interminable and what we have been calling the chronic.

In response to the question of when an analysis might come to an end, Freud writes what might be one of his most brilliant sentences and certainly his most definitive on the problem of terminality: "The analysis is over when the analyst and the patient stop meeting for analytical sessions."[28] Perfection. It is true that the rest of the essay stresses the contingent nature of ends and cures and whether permanent healing is possible or whether analysis can provide a certain preventive treatment to future illnesses. And to conclude, Freud argues that it is the analyst who should return to analysis every five years, an interminable practice as a way to better engage the realities of counter-transference. The indispensability (and ambiguity) of the "and" between terminable and interminable, therefore, is similar to the "and" between analyst and patient—roles that are never so discrete, the one necessarily supplemented by the other. But there is something about the initial tautology that hints at something more, something about how psychoanalysis is not possible without the sessions, without the meetings (in real time and over time) of the analyst and patient.

We can understand this imperative as a way to privilege the problem of praxis—perhaps returning us to how Freud criticized the young doctor for knowing nothing about the practice of psychoanalysis, a criticism leveled regardless of the accuracy of the young doctor's interpretation of the divorced woman's symptoms. Although regularly understood as a synonym for practice, I want to argue for a different understanding of praxis: as the name for the desire to unite theory and practice, however utopian such a project is fated to be. Praxis denotes the ceaseless movement between thinking, understanding, experimenting, acting, and changing. None of these categories exist autonomously, but are always supplemented by each other. Indeed, this tension—the irreducible gaps that structure these categories—is as inescapable in the clinic for the psychoanalyst as it is in the studio for the artist, and on the street for the political philosopher. From the very beginning, psychoanalysis has factored into its discourse this inescapable problem of praxis, thus leading to the almost universal principle that every analyst must be analyzed. This principle, however, is not only about training more capable analysts, but (and Freud was, indeed, the first to recognize this) about engaging more effectively the very problems of psychoanalytic theory. We can go one step further and argue that when an analyst is engaging a patient, he or she *is* theorizing—and the particular, long-term relationship between analyst and patient (in which their very relationship itself becomes the practical material to be analyzed, or in this case theorized) is an ideal place to engage this principle.

Here we might want to make a play on one of the other great statements about praxis, Marx's eleventh thesis on Feuerbach: "The psychoanalysts have only interpreted the subject; the point, however, is to change it." But, like Freud, when Marx invokes philosophers and the world he is not resorting to an easy argument against ivory-tower academics while celebrating action-oriented world change. Rather, Marx puts into question the relation between interpretation and change. In fact, Marx does not imply that interpretation or philosophy is the culprit, only that interpretation and philosophy that does not change the world is inadequate—and many times functions to strengthen the here-and-now power structures regardless of how militantly such theory purports to support the need for radical transformation. The opposition, therefore, is not between interpretation and change, but between any action (be it theoretical or otherwise) that reproduces the status quo and any action that changes it. It is for this reason that Freud struggled with the issue of psychoanalytical rules and protocols —for rules and protocols are best determined in relation to, and are contingent upon, the effects of each individual treatment.

But what about the lack of Freud's own analysis, a deficiency he tried to overcome by analyzing his own dreams (leading to the publication of *The*

Interpretation of Dreams [1899], his most monumental work, and one in which approximately 30 percent of the more than 140 dreams analyzed are Freud's own)?[29] What of Freud's long correspondence with Wilhelm Fliess, his close friend and confidant who served as key interlocutor (primarily via the exchange of letters) during Freud's early work (1887–1902)? Indeed, the combination of *The Interpretation of Dreams* and the Fliess correspondence is effectively Freud's self-analysis, as indicated in the way Freud implicates Fliess in the section of the following letter, written in July 1897: "I still do not know what has been happening in me. Something from the deepest depths of my own neurosis has ranged itself against many advances in an understanding of the neuroses and you have somehow been involved in it. For my writing-paralysis seems to me designed to hinder our communications. I have no guarantees of this; they are only feelings of a highly obscure nature. Has nothing of the kind happened to you?"[30]

And the following month Freud wrote, "Things are fermenting in me, but I have finished nothing. I am well satisfied with the psychology: I am tormented with grave doubts about my theory of the neuroses. . . . [T]he chief patient I am concerned with is myself. . . . The analysis is more difficult than any other. It, too, is what paralyzes my psychical strength for describing and communicating what I have achieved so far. But I think it must be done, and is a necessary intermediate stage of my work."[31]

Freud recognizes that his own abilities to "describe" and "communicate" depend upon the success of his own analysis. There is a practical limit to what is possible to think and express. There is no theory without practice, no practice without theory. And, perhaps most significantly, there is no resolution to the theory-practice problem. But this impossibility does not mean that all engagements with the problem are equal. In this case, we might say that there are better and worse failures. And two of Freud's most productive failures were his formal attempt, by way of his very sentences and stylistic choices, to overcome this impossibility, and his way of recognizing that the problem of praxis itself has a history and, therefore, shifts from one historical moment to the next, thus shifting what constitutes its most effective engagements.

The form of Freud's own writing is often considered his most lasting legacy. The case studies are fraught with literary flair and look nothing like the prevailing discourses of the day (be they in the medical sciences or in the humanities). It is often said that Freud was halfway between poetry and science, which is one of the reasons he was awarded the Goethe Prize in 1930. Even so, what is most interesting about Freud's writing is not that he produced a singular style required by the new discipline called psychoanalysis, but that he gave himself over to the inescapable fact that no matter what form he chose (from the scientific to the poetic) it turned out utterly farcical. If he is too scientific and sure of himself (as his earlier work is) then he reads like a terrible reduction of the human condition. If he is too poetic and ambivalent then he comes across as someone truly unfit to make clinical claims about the very real suffering of his patients. Indeed, this is the point: there is no way to properly represent psychoanalysis. And this is its greatest strength—one that productively pressurizes Freud's entire corpus.

This is probably the most important similarity that Freud shares with Marx. Both thinkers resist the systematization of their thought by the way they anticipate the reader in their prose. There is nothing that betrays Marx's work more than so many tired diagrams of how the base relates to the superstructure. This is not to suggest that the various formulas and schematics that Marx provides are not central to his work, but one cannot abstract these fundamentals from the work without attending to how they emerge within the writing itself. *Capital* is the great example of this.

Marx writes and organizes *Capital* so that it must be read (or "experienced" to use an old-fashioned term) more than understood. By this I mean that the pedagogical and performative dimensions to the text are what are most compelling. The moment one jumps over these dimensions is the moment one is set on the slippery slope of dogmatism and orthodoxy. *Capital* begins with a meticulous analysis of the commodity, the simple commodity. In the very first line of *Capital*, Marx writes, "The wealth of those societies in which the capitalist mode of production prevails, presents itself as 'an immense accumulation of commodities,' its unit being a single commodity. Our investigation must therefore begin with the analysis of a commodity."[32] One must begin with the particular, with the detail, and rise to the abstract, the general. Marx feared, however, that such an arduous concentration on the detail at the very beginning of *Capital* might put off his readers, leading him to begin with a note of encouragement in the French edition. "That is a disadvantage I am powerless to overcome, unless it be by forewarning and forearming those readers who zealously seek the truth. There is no royal road to science, and only those who do not dread the fatiguing climb of its steep paths have a chance of gaining its luminous summits."[33]

One must actually read *Capital* (in the same way that one must read a novel as opposed to simply know its plot in advance), an act that implicates the reader in the system, in the history the text is describing, and in the very mode of the text's time. One must give oneself over to the small stuff in order to engage the larger structure as something more than so many predictable answers to so many predictable questions. The reader is inspired to experience the tension of how capitalist society is—at once—both personal and impersonal. It is an impersonal system that operates by a certain logic (the very logic that is inscribed within its most basic detail, the commodity), and it is an association of interrelated individuals, all of whom are ineluctably integral to and personally invested in (however much against their interests) the system's reproduction. This tension runs throughout the work, as Marx is careful not to moralize, not to account for inequality and injustice by pointing fingers at corrupt individuals.[34]

All the same, Marx cannot hold back. He is ruthless, sarcastic, and acerbic when calling out the opportunistic behavior of the bourgeoisie. The paradox of how capitalism functions in such an impersonal and personal manner is performed on the very level of Marx's stylistic choices—the way sparks fly when he brings together two opposing and incompatible forces at once. The same can be said for the way Marx overloads *Capital* with a combination of pure theory, historical analysis, and journalistic reporting. In fact it was precisely for this excessive mix of methodologies that the great Japanese political economist Uno Kozo criticized *Capital*.[35] Uno believed that one must separate the three levels so as not to compromise the rigor required to develop each one. Fair enough. But Marx's methodological mistake is at the same time a formal solution to an impossible ideal: one cannot thread the methodological needle without drawing blood. And it is precisely this blood that energizes the work and turns it into something more than another safe academic exercise or futile volunteeristic rant. Marx's compulsion to give more than what was required, his "academic error," provides a model for the reader who is also required to give more—to be both extragenerous and extracritical of the text, which can only occur when one engages and reflects in a way opened up by the dialectical method itself.

This is another way of arguing that capitalism, as a system, resists representation. Capitalism (like the unconscious for Freud) is an abstraction, a set of relations, a system. It is nowhere to be found. One cannot hold it and measure it and prove it. But the stuff of capitalism is everywhere and readily accessible to study. All of the commodities, the bank notes, the heavy lifting, the extreme consumer desire, the prized philanthropic acts, the neuroses (from hypersensitivity to disassociation), are not capitalism —they are the effects of capitalism. To separate the various levels of capitalism and analyze them systematically and rigorously, as Uno would have it, is in fact to misrepresent what capitalism is, because capitalism does not honor discrete levels, nor is it systematic in some one-dimensional, mechanistic way. Capitalism is a system, but one that can only be flashed by way of its most fundamental detail.

Freud, too, believed that the detail was the most effective starting point. It is the detail as parapraxis (one of those unintended slips) within the analytical session that opens the backdoor to the most meaningful realizations. But what about the reader of the Freudian text? Unlike the analysand who always begins a session, the reader is always responding to the psychoanalytical sentence—is always betraying the most basic rule by going second. And this leaves Freud the writer in an equally compromised position, as he is forced to begin. Freud was stuck. How could he represent psychoanalysis in a mode that was so unpsychoanalytic? Especially, since (like Marx) he did not see his audience as solely made up of professionals. How could he incorporate the reader into the prose in a way that was both personal and impersonal and that generated more complicit immediacy than empathic distance, while at the same time generating more analytic distance than distracting immediacy? Ultimately, Freud was searching for a mode that would not generate the same type of resistance he was witnessing in his patients.

The bind here is that the moment one directly describes a psychoanalytic session, theory, or concern is the very moment the farce begins. Those who have been through an analysis often share the experience of sounding ridiculous (to themselves especially) when describing the experience to others. Four days a week. On the couch. How could this come across as anything but absurd? Especially when we remember that most who choose to undergo psychoanalysis are not suffering from severe mental health concerns. Here we might want to return to the psychoanalytic anecdote I told at the beginning of this book: "In that case I better see him right away." This anecdote does not only indicate that psychoanalysis lends itself more toward a radical intervention in the chronic than in the acute, but that the whistleblowing "aha moments" are less significant than a more comprehensive system overhaul. Within this more radical paradigm, the chronic (the everyday, non-urgent crises that are internal to the stability of the patient) is treated as acute, while to manage the symptoms (to treat them as chronic) is to terminate any possibility of significant change. The time for intervention, in other words, is precisely the time when the patient is "best managing" his or her symptoms. To retain cure as a guiding principle (as a strategic and formal destination rather than an impossible content), therefore, requires that the patient terminate his management routine and dependence on chronic time.

PSYCHOAN . . . , FILM THEO . . . , REAL TIME

This subtle attention to time accounts for why film and television almost always fail when representing psychoanalysis. Even Hitchcock fails. In *Vertigo*, when Scottie (played by Jimmy Stewart) finds himself nearly catatonic in the psychiatric ward after contributing to another death due to his acrophobia, the head psychiatrist responds to a question by Scottie's longtime and long-suffering friend Midge (played to perfection by Barbara Bel Geddes) about how long it will take to pull Scottie out of his traumatic state. "Well, it's hard to say," the doctor begins, "at least six months, perhaps a year. It really could depend on him." Midge responds, "He won't talk." The doctor explains, "No, he's suffering from acute melancholia together with a guilt complex. He blames himself for what happened to the woman, we know little of what went on before." Midge replies, "I can give you one thing, he was in love with her." The doctor adds, "Oh, that does complicate the problem, doesn't it," to which Midge rejoins, "And I can give you another complication, he still is."

Dejected, head-down, and fighting back tears, Midge finishes with the doctor and slowly walks down the hospital's long corridor. Hitchcock takes his time with the shot and sets it against the deep cello notes of the score and the monochromatic blueness of Midge's coat matching the ward's walls. One cannot help but find the melodrama of the scene and the use of psychoanalytical language ridiculous—Hitchcock's mockery and self-mockery—especially in comparison to how subtle and brilliant *Vertigo* is overall. Hitchcock seems to know that there is no way to directly represent psychoanalysis, however much his entire oeuvre is one of the greatest engagements with psychoanalytic concerns. Still, by including ridiculous scenes of psychoanalysts and psychoanalytic dialogue (perhaps the most ridiculous lines are those of Gregory Peck as a psychoanalyst in *Spellbound*) Hitchcock actually makes the films stronger. This is because Hitchcock's films foreground the fundamental antinomy of theory and practice —that neither of the terms can properly represent the other and that there is no overcoming the theory-practice problem itself. Still, this resolution must remain for us as an ideal (or at least a placeholder) in order to energize both our theorizing and our practicing. Hitchcock's films are strongest when they do not directly refer to psychoanalysis. In fact, we can go a step further and argue that psychoanalytic clinical practice is strongest when it does not directly refer to psychoanalysis.[36]

This inability to directly represent psychoanalysis is most brilliantly figured by Hitchcock in *Vertigo* when Gavin Elster (the old friend of the Jimmy Stewart character, and the one who is behind the deceit) asks, "Scottie, do you believe that someone out of the past, someone dead, can enter and take possession of a living being?" "No," Scottie responds. Elster goes on, "What would you say if I said this has happened to my wife?" "Well," Scottie replies, "I would say to take her to the nearest psychiatrist, or psychologist, or neurologist, or psychoan . . . or maybe just a plain family doctor . . . I'd have him check on you too." Scottie cannot pronounce the whole word "psychoanalysis," nor can Hitchcock wholly show it on screen.

I will return to representations of psychoanalysis, but now I want to ask how psychoanalysts represent cinema. The short answer: not very interestingly.[37] This is to say that the focus is almost always on how psychoanalytical themes are represented in film, not, for example, how psychoanalysis and film share similar theoretical and practical problems—most notably the problem of time itself. As a test case let's take Michael Haneke's film *Caché* (2005)—a film that tempts us to psychoanalyze its characters and content at the risk of burying the more radical elements of the film. At the very core of *Caché* is precisely this incommensurability of scanning the film for psychoanalytic themes and analyzing the very temporal relationship between film and psychoanalysis. Falling back into representations of psychoanalytical problems on the level of the film's content invariably crowds out other modes of analysis. The search for psychoanalytic themes is tempting: the bourgeois protagonist's inability to connect with others

The opening static shot from the film *Caché* (2005), directed by Michael Haneke.

seems to stem from a repressed childhood memory of poorly treating an Algerian boy, as well as from a repressed social memory of an incident on October 17, 1961, when 200 Algerian protesters died at the hands of the Paris police. It is the impossibility of holding both forms of analysis that is foregrounded in the film.

Caché begins with a long-take video image of the front of the Paris home of Georges and Anne (and their twelve-year-old son, Pierrot), shot from across the street. For about two and a half minutes the image remains with no dialogue. All the viewer sees is the home as pedestrians, cyclists, and automobiles move across the frame. For the shot's first minute and a half, *Caché*'s titles are typed on the frame. Soon we learn that Georges and Anne are watching the videotape that we are watching in their home, the same home that has been represented in the surveillance footage. This shot sets up the film's narrative, as Georges and Anne attempt to figure out who is sending the videos (followed by equally provocative pictures and post-cards) and what they mean.

This opening shot is at once double, requiring two different forms of viewing. The video footage is the opening shot of Haneke's 35-millimeter film *Caché*, demanding viewing strategies that must piece together this narrative information with what will follow. Like Georges and Anne, we too are asking what this means, who sent it, and how it will shape what is to come. Although the long take is shot on video, it is in fact a video image

The closing static shot from *Caché*, including Pierrot and Majid's son (foreground left) in front of Pierrot's school.

within the larger 35-millimeter film, a fact that is confirmed by both the opening title sequence and the shift to a differently textured image when inside the home. This first viewing strategy is the dominant one of narrative cinema and not unlike the interpretive strategy of what we might call narrative psychoanalysis, the garden variety kind that lends itself most easily to popular representations. It is centered on meaning-making, on chasing down every sign and ordering these signs so as to make sense of the story. Meaning transcends the single shot.

The second viewing strategy provoked by the opening shot is one less connected to narrative cinema and more to reality television images. It requires a way of viewing that is less interested in how the shot relates to the larger narrative than in how the banal images might explode at any moment, an explosion that will come out of nowhere and relate to nothing save our own beating heart. Meaning is immanent in the single shot.

The last scene of Haneke's film seems to engage the incommensurability of these two ways of viewing. Again, a video image returns, shot at a medium distance, with the camera resting on a tripod. This time the camera is directed toward the steps leading to a school Pierrot attends. For a minute or so, we watch the everyday scene of kids getting out of school, chatting with each other, looking for their bicycles, and getting into their parents' cars. The camera does not emphasize any part of the frame, nor do the diegetic sounds. On the left side of the frame, in the middle of the banal school scene, is Pierrot speaking to an older boy (who is in fact the son of Majid, the grown man whom Georges betrayed as a young boy). We do not hear what the two characters are talking about, nor can we decode much, save the fact that they know each other and seem to get along. What is extraordinary about this image is that it is quite likely that the viewer does not even notice the presence of Pierrot and Majid's son.

If the presence of the two characters is established, then it is hard to resist filling in the blanks of the story. The moment the final shot is "straightened out," in other words, this new information is retrofitted into the film's larger narrative. Significantly, this returns us to a symptomatic mode of interpretation, notwithstanding that this mode is but one of various interpretive strategies. *Caché*'s argument reawakens the utopian desire (however impossible) of holding together incommensurable forms. It is that moment, right after we realize that the boys are in the shot, and right before we recuperate the new information as central to our theatrical understanding of the film, which represents the film's utopian dimension—a dimension that does not overcome the antinomy, but centers it before forcing us to betray it.

How do we hold a specific moment—and a specific mode of time—without managing it to death? There is one more scene in *Caché* that exquisitely resolves this seeming impossibility. Pierrot and others on his swim team are practicing underwater turns on one side of the school pool. The team coach is also in the pool, watching carefully and commenting about technique to each member after his turn. Once one boy is in the middle of making his turn, another boy starts a turn, while a third boy just emerges from underwater. Each boy follows the boy in front of him by about five seconds. Since there is a delay before the coach's voice can be heard, he gives advice to the boy emerging from one turn while looking at another boy making his turn underwater. A looping of time is soon established in which past, present, and future overlap, giving the scene an acuteness that seems to break out of the film's narrative altogether. While this time chamber is cranking up, Pierrot's parents (still traumatized by the surveillance videos and their own failing marriage) look on from the stands. There are other spectators as well, some videotaping the swimming practice. Here, again, we have film in competition with video, each representing a different mode of interpretation, experience, and time. In this case, film represents the chronic, the management of crisis by way of linear narrativity (the sharpness and depth of estranged feelings in Georges and

Another film still from *Caché*, capturing one of Pierrot's swim team practices.

Anne, the tension building around their intruder), while video represents the terminal, the exploitation of crisis by way of a forced event (the flatness of the video image that is impersonal and destructive). Each of these modes, despite their radical potential, is effectively reactionary, and each ends up recuperating the existing discourse of power. The looping of time in the pool, however, interrupts both modes and flashes (like a crystal just before being crushed) the radical possibility inherent in cinema, psychoanalysis, and politics.

This revolutionary possibility is internal to these three discourses. No matter how susceptible each is to chronic time, to the snuffing out of change so radical that it is unrepresentable at any moment before it erupts, what persists in cinema, psychoanalysis, and politics is the possibility of their very ends, of their very deaths. The death of each would not mean the ultimate end of moving images, or the end of clinically engaging or theorizing human behavior, or the end of investigating and participating in the social relations of power. Rather, the death of cinema, psychoanalysis, and politics simply means that a fundamental change in the historically specific expressions of these forms occurs. And this change is so primary that it effectively changes the very categories themselves. Not only is such a radical change—such a death—unimaginable and unrepresentable before it occurs, but it is also unimaginable and unrepresentable after. This is because a new expression will not seem new at all, nor will it seem already on the brink of death. Moreover, this process is never linear, there are flashes of each form's future death within its present life. *Caché*'s looping of time in the pool is just such a death flash. I will develop this analysis of death and its relation to life in terms of continuity and discontinuity in this book's third part, on the already dead; for now we move ahead with the problem of representing psychoanalysis.

The principal reason why psychoanalysis is so difficult to represent cinematically is the fact that the time of psychoanalysis is positively non-cinematic, not to mention how noncinematic the banality of psychoanalysis is. In other words, to dramatize analysis, to submit it to a cause-and-effect logic of symptom relief, is to immediately betray it. The time of analysis cannot be shortened, just as the time of filming analysis cannot be shortened. Much can be done, however, in the real-time of analysis and film.

The HBO series *In Treatment* engages this limit by organizing its fictional episodes around the weekly sessions of a psychotherapist, Dr. Paul Weston. Paul treats several patients during the week. Each episode of the series focuses on one session, so that the audience follows the same patient (or couple or family) on the day they see Paul. Fridays are reserved for Paul's own sessions with Gina, his therapist. Almost every episode (thirty-five in the first season) remains in Paul's office as he negotiates the neuroses, separations, panic attacks, affairs, suicide attempts, transferences, and a DSM list of conditions one would expect to find in his mostly middle-class East Coast patients. Each episode usually gives the impression that we are following a fifty-minute session—beginning at the beginning ("How did the week go?") and ending at the end ("See you next week"). But each episode is only approximately thirty minutes. Where do those extra twenty minutes go? And, coming at this from the other direction, why are therapeutic sessions usually fifty minutes?

The fifty-minute session has been standard since Freud, and, although it has been challenged at various moments over the past ninety years (most notably by Lacan), it is still the dominant duration of a single psychoanalytic session. The hour is a functional interval. Properly coordinated with the time of capitalist work, the fifty-minute session grants to the analyst the needed ten-minute rest-period to write notes, take a call, or go to the washroom, not to mention that it allows analysands not to pass each other on the way to and from a session. The hour is also the key unit of labor valuation within capitalism—labor time being the most important commodity bought and sold by capitalists and workers. This practical dimension of time has often been conflated with the methodological principle of the fifty-minute hour, leading some to argue that the analysand needs enough time to free-associate or to manipulate the time (as when an analysand might wait until the last five minutes of a session to drop a bombshell). As for not going over an hour even when the dialogue might be at a crucial spot, some argue that the rigid and arbitrary termination encourages the analysand to hold onto tension and frustration, thus practicing the healthy act of letting time pass before returning to a burning issue.

Why are television episodes organized around a similar block of time, either thirty or sixty minutes? Richard Dienst analyzes the relation between television and time by appealing to Lacan's famous appearance on French television in 1974. Lacan, looking directly into the camera, began by stating, "I always speak the truth: not all of it, because there's no way to say it all. Saying it all is materially impossible: the words are lacking. It's even through this impossibility that the truth holds on to the real." Dienst argues the following: "Like Lacan, television cannot say everything. Although it always looks as if it is trying, as if it is about to get everything out. Lacan's gesture . . . summons the whole by admitting the impossibility of articulating it." Although television cannot say or show everything, it can, according to Dienst, represent time in seemingly contradictory ways that define contemporary life—still time and automatic time. "Televisual stills are created by switching away from a picture, pushing past one toward another, by halting a movement or adding a different one. And these stills do not add up or follow one another: each turns over and disappears from view. Meanwhile, automatic time appears when an image is switched on and left running, so that it is no longer an image *of something*: it is the time of the camera's relentless stare, persisting beyond the movements of objects and scenery that pass before it. If still time slices off images and designates them as past, then automatic time opens onto an anticipated future: it is an image waiting for its event to happen."[38]

Dienst writes these sentences in 1994, five years before the first season of *The Sopranos* and between the two Gulf wars. Although CNN existed, video streaming on the computer and the new dominance of cable television programming had yet to hit a critical mass. This final moment of broadcast television's supremacy might account for Dienst's argument at the end of *Still Time in Real Life*: "Neither still nor automatic time can reach a final end point or fulfillment, where each would finally become the other."[39] But in today's world, in which the chronic and the terminal are reconfiguring (with the chronic becoming dominant), still and automatic time seem to have merged, as the short term and long term have collapsed. Just as the new chronic manages the short-term crisis by extending it into the long term, so does it manage the long-term crisis by turning it into a problem of the short term. For example, one can watch a whole season of *In Treatment* over a period of a few days, either by bit torrenting or streaming each episode in succession or by marathon-viewing the DVD set. This viewing practice turns the long-term continuous season into a single discontinuous image. While, at the same time, watching the whole season over a short-ened overall amount of time turns a discontinuous season of episodes into a short-term, continuous viewing experience.

Cable arc narratives enable new viewing experiences, ones not shaped by the weekly and seasonal time of TV. Another example of this comes by way of Michael Apted's *UP* series, in which he started filming the lives of fourteen seven-year-old kids in 1964 and has continued to make install-ments about their lives every seven years up until the present. One can rent the DVD set and watch, in succession, seven films until reaching *49 UP*, made in 2005. It is one thing to view the next installment of these docu-mentaries every seven years, but it is quite another to view them at one time. It is like time-lapse photography, or time-lapse life. The point is not that there is a new form of representing the passing of time, but that there is a new form of viewing and experiencing the passing of time—one that I am suggesting is not simply shaped by television and popular culture but by a much more encompassing temporality affecting every aspect of our lives. The temporality of a series (even one that, like *UP*, covers decades and the lives of a generation of people) finally cannot escape viewing practices that drag it into a perpetual, unremarkable present, just as the hype over tar-geted drug therapy effectively snuffs out the desire for more radical medi-cal and social practices. This temporality—this new chronic—is most per-fectly coordinated with and challenged by that of palliative care.

PALLIATIVE TIME

At the very beginning of the *Oxford Textbook of Palliative Care*, the recog-nized authoritative text on the topic, the editors explain why the term "palliative care" is used instead of "terminal care" or "hospice care."[40] First, there is an ambiguity about what constitutes the terminal and when the terminal should be invoked. "Does it refer to the last hours, days, weeks, or even months of life?" the editors ask. Second, the "terminal suggests that all is finished, that there is neither the time nor the opportunity to do more, and that active treatment is unjustified and might well be undignified." The editors go on to argue that " 'Terminal Care' suggests a preoccupation with dying and death, grief, loss, and sadness and, although we do not wish to create a sanitizing definition, the care outlined in this book affirms life rather than death." As for hospice care, there is a perception (especially in the United States) that "it is a soft product without a scientific base"— a fraudulent new age alternative that gives up too soon on medical advances. This leaves us with the least ambiguous term, "palliative." The World Health Organization defines palliative medicine as "the active total care of patients whose disease is not responsive to curative treatment. . . . Palliative care . . . affirms life and regards dying as a normal process . . . [it] neither hastens nor postpones death . . . provides relief from pain and other distressing symptoms . . . [and] integrates the psychological and the spiritual aspects of care. . . . Control of pain, of other symptoms, and of psychological, social and spiritual problems, is paramount."[41]

Palliative medicine has traditionally been one of the more marginalized and least resourced fields within medical care. The dying were generally given the lowest medical priority, as death was understood as a "medical defeat or statistical embarrassment."[42] Before the discovery of anesthetics (which enabled safe and effective invasive surgery) and the widespread use of antibiotics in the 1950s (which "cured" common infections), however, all medicine was palliative. This means that the paradigm shift from palliative to curative within medical practice occurred over the past fifty years. Also contributing to this shift were the priorities of medical education. Effective communication skills and training in medical ethics, psychological concerns, political and cultural influences, as well as in spiritual questions, were rarely granted more than token space in the dominant medical curricula. By 1992, an average of less than thirteen hours were devoted to such concerns in a five-year course.[43]

However, as we saw with television and the collapsing of its two opposing temporalities ("still" and "automatic" times), the distinction here between two paradigms of medical care is today becoming nearly unrecognizable. With the new chronic paradigm emerging, the threshold that separates palliative from curative medicine becomes less defined. On the one hand, palliative care is marginalized as pharmaceutical executives, celebrity doctors, and scientists search for the silver bullets to conquer cancer, those discoveries that are readily sensationalized by the media and sustained by so many cancer ribbons and charity events. On the other hand, a new version of palliative care is returning as the paradigm of disease management becomes dominant. The time of the palliative, therefore, has changed—it no longer has a time limit and now refers both to care given in the last few days of life and care given over an entire lifetime.

With this simultaneous marginalization and recentering of the palliative (with both tendencies submitting to the repressive dimension of the new chronic) also comes new and radical possibilities for the palliative—ones that trigger the creative and revolutionary force of the terminal and that reopen the future as something more than a mere linear extension of the present. With the radical time of the palliative released, moreover, new ways of caring for and investing in the present open. For example, new ways of caring for the dying that refuse to understand death as defeat or as an economically unproductive stage of life (from elder care to palliative care experiments) can also be applied to ecological, economic, political, and cultural conditions.[44] The way that we care for our species, the nation-state, or for the planet, for instance, would no longer be limited by reactionary fantasies of salvation or cynical strategies of sustainability, but the palliative present would be cared for and opened up by creative engagement—a series of noninstrumental acts that could very well turn out to be the royal road to necessary change. A radicalized palliative model that reconfigures the time of living and dying as well as the chronic and the terminal might turn out to be the most productive model for engaging the acute challenges of the present.

Instead of this possibility, however, what is becoming clear in the realm of cutting-edge medicine is a profound shift in which the more reactionary dimension of the chronic is overtaking the more radical dimension of the terminal—with all of the associated temporal implications. There is the refunctioning of two key temporal categories: "crisis" and "the meantime." The meaning of "crisis" is no longer based on the word's Greek etymology, in which *krisis* refers to an inescapable decision, a turning, that must be made immediately. But now crisis is extended, rolled out flat all the way to the indefinite "long term." And there is the meantime, now permanent, thus permitting the present to fully colonize the future.

The Political

When we move to the political today, we see a chronic form of politics becoming dominant, one that turns on a rethinking of revolution. Today the category of revolution (like the medical category of cure) is losing its force. Revolution as wholesale change, as taking over the state, as acting for a radically alternative future, is called into question. The relation between the medical and political today is usually pursued from one of two directions. First, through the category of biopower, the Foucauldian concept that brings together population control with modern regulations on the human body and the production of life. And second, through a straight political economy of life in terms of everything from the profits and losses incurred by pharmaceutical corporations to intellectual copyright concerns over scientific patents. I want, however, to explore a third relation that examines how the recent transformations of the category of political revolution share crucial formal similarities with the transformations of the medical category of cure (namely, the dominant mode of time) and that the way we understand these similarities (the way we discard or retain them, deconstruct or dialectically crank them up) has significant implications.

The classic way by which revolution has been engaged within political theory is through the problem of revolution's relation to reform. The radical position has always emphasized revolution while the social-democratic and liberal position has emphasized reform. The revolutionary calls into question the reformer's understanding of systemic limitations and argues that the reformer's tinkering along a structure's edges only reproduces and, more suspiciously, justifies the very system under critique. The reformer calls into question the revolutionary's emphasis on "big-picture" commitments while caring little for the details of political action and the small changes to quality of life that reform can enable.

In the mid-1990s, Fredric Jameson presented the lecture "Actually Existing Marxism," in which he argued that the most important moments of revisionism (or reform) of the revolutionary problematic emerge at moments when capitalism itself undergoes structural metamorphoses, beginning with Eduard Bernstein at the turn of the nineteenth century and leading up to the various post-Marxisms of the 1980s.[45] But, and in elegant dialectical fashion, Jameson argued that if capitalism still exists then so does a revolutionary Marxism, for Marxism is simply the analysis of capitalism.[46] If this were ten years ago, I might spend the remainder of this section arguing not only why capitalism still exists (why the fundamental rules of capitalism are still operative, namely, crisis, surplus value, and class conflict), but also how the system itself has transformed into a different stage, a stage that these days goes by the unanimous, however questionable, name globalization. But, no doubt, we have heard this story (of similarity and difference) several times before.[47] Instead of rearticulating this relation between Marxism and capitalism (their historiographical inextricability, their mutually constituting dynamic), it is high time to make a move on this debate and argue the following: that the most important moments of a return to a revolutionary politics *follow* the radical expansion of the system, when the trauma of capitalist crisis is temporarily recuperated—when capitalism is most banal, most visible, most unmediated, most true to its fundamental logic is when politics itself is most visible, unmediated, and logical. Or to put this another way: the more vulgar capitalism becomes, the more vulgar politics must become (the word "vulgar" in this context should be understood in the most nonpejorative sense and in a way that is distinguished from what goes by the derogatory nineteenth- and twentieth-century name "vulgar Marxism").

THE NEW VULGAR

We can understand the current state of revolutionary thinking and practice as a symptom of this historical present—a moment when to ask and answer the question of what it means to "be" revolutionary is a task not only for the sad militant or old-timer, but a focus of study for a whole new generation of students, activists, workers, intellectuals, artists, and, lest we forget, the poor and disfranchised the world over, who now have more reason than ever to look toward a reconfigured revolutionary movement for some sort of analytical framework and political project. I have now set up two tasks: first, to elaborate this contemporary moment of vulgar capitalism (in which the truth content of the economic logic is as close to the bone as it has ever been) and to elaborate what a revolutionary politics might be today and, second, to elaborate how this "new vulgar" relates to the shifts in dominant temporality that I have been calling the new chronic. The first task is an intellectual and political project that is less about economic determinism or a bald reflectionism in which the aesthetic is at the mercy of some tight-fisted totality, and more about how the current situation has opened up new configurations and new possibilities that are quite specific to our own day—possibilities and configurations that require a reemphasis of the political-economic without forgetting the lessons learned from a half century of critical theory's attention to ideology, aesthetics, subjectivity, and what more generally comes under the name "the cultural turn."

When mentioning vulgar capitalism, I am not only referring to the logic and effects of capitalism the world over, but also to the current hegemonic ideology of capitalism shared by many on both the left and the right. During the Cold War and the immediate post–Cold War moment, over-stuffed and overcoded political categories such as freedom, liberty, and democracy functioned to bully many into acquiescence, so that any criti-cal questioning of their own sacrifices was deemed unpatriotic. In other words, injustice was explained away by invoking political necessity. While today economic categories such as profit, sustainability, and expansion have usurped the ideological purchase of the former categories. What marks our present post post–Cold War moment is an apology for in-equality based on the unapologetic logic of the capitalist market—one that is not at liberty to suspend the rule of profit and expansion under any circumstances.

What is peculiar about this recentering of the economic is that it has not quite taken hold within much of the cutting-edge scholarship occurring in the humanities and social sciences today. For example, I am thinking about some of the work coming out of political science, history, literature, and philosophy that focuses on such categories as human rights, sovereignty, governmentality, and the biopolitical. A certain bracketing of capitalism, and the economic more generally, usually results from stressing the dia-chronic development of these categories (from the premodern to the pres-ent) so that, for example, the problem of human rights becomes a consis-tent problem that persists from one historical moment to the next, rather than becoming a problem that takes on qualitatively different meanings and effects depending on the logic of the social formation in question. One reason why the economic has become the missing term today is due to a backlash against the repeated overemphasis of the economic as the first and final cause of everything, the cause of philosophical ideas, political institutions, aesthetic experiments, even our most private human desires. And this takes us to the problem of vulgar Marxism and how we might be able to rethink this term for the present moment.

We begin with two of the most important features of vulgar Marxism, namely, its historical progressivism and economism. As for Marxism's past reputation of being "teleological," and present claims that it still is hostage to writing history in advance, thus snuffing out any contingency, let alone human agency, we only need to look as far as the Frankfurt school. Rather than tease out a philosophy of history in the work of Theodor Adorno, Herbert Marcuse, or Ernst Bloch (not to mention the most obvious member when talking about a philosophy of history, Walter Benjamin), I briefly turn to their aesthetics to argue that we can read in the general principles of this work a criticism and reworking of an all-too-clean diachronic movement from capitalism to communism.

In the most general way, when that thing known as critical theory engages aesthetics it is interested in how formal invention attempts to come to terms with the most pressing sociopolitical concerns of a particular historical moment. What cannot be figured in, say, a social or political language, can be glimpsed by way of the aesthetic (sometimes in advance of the sociopolitical itself). An experiment with literary tense (or genre) might escape the orthodoxy of a literary structure and thus figure a political escape from a particular historical moment, an escape that would be impossible by any other means, save radical social transformation. The point here is that a particular aesthetic category "means" only in relation to its social context. It is for this reason, incidentally, that postmodernism is more productive primarily as a historical category, rather than a stylistic one, for when it is only understood as style (composed of a nonlinear narrative or composed by an unstable narrator) we can find similar examples throughout the history of modern and classical culture, not to mention throughout every national culture. The same aesthetic technique, therefore, works differently at different moments. Already we can begin to detect a method of analysis here that is less about a predictable march through history and more about the way history structures the unpredictable, the unthinkable—how thought, even the thought about this very method, is contingent on the historical moment during which it is enacted.

For a contemporary example of how such a conceptualization might matter today, there is the category of the nation. One of the most unfortunate and unproductive developments in the debate over globalization has been the question of the relative strength or weakness of the nation-state. It seems that every discussion about globalization invariably returns to an argument about whether the nation still matters, whether national leaders, as well as resistance groups, can organize around the nation or whether any reliance on the nation is anachronistic—a failed strategy in advance. Is the age of the nation-state over or is the nation as strong and central as it ever was? To protect against falling into the false problem of this debate, we might want to put to service the method we just established. Which is to say that of course the nation-state still matters, of course neoliberalism absolutely depends on a strong nation-state system to enforce its global policies, of course nationalisms and national identities are as strong as ever, but all of these instances in which the persistence of the nation is clear exist within a different set of structural relations. It is not a case, therefore, of the nation being more or less of anything—just as "postmodern" literature is not about more or less stylistic consistency or some "new innovation" like an untrustworthy narrator. Rather, the nation and nationalisms function differently than they did at earlier moments, just as the same formal strategies we can identify in works from earlier periods function differently in the "postmodern" period.

Within the globalization debate, the nation, therefore, functions as a red herring, diverting our attention from what is happening in the larger capitalist system. And this, finally, takes us to Marxism's sore thumb—its economism. Economism was first introduced as a "bad term" by Lenin in 1899, when he criticized some groups in the Russian social democratic movement for separating political and economic struggles and concentrating their efforts on the economic.[48] For Lenin, therefore, economism became a problem of practical politics, while as a more theoretical term it refers to the determinations, by the economic base, of social life as a whole. So what to do with the economic today?

In order to understand the omission of the economic in North America we can begin with the dominance of neoclassical economics within university economics departments and the shift away from political economy to political science. In other words, it is not just about the continued backlash against Stalinism or economic determinism (which perhaps better explains the European situation, since Stalinism, as a theoretical paradigm, never had a strong effect on North American leftism and was, even in the Stalinist period, a contentious issue at best within the Communist Party of the United States of America). Rather, the omission of the economic relates to the discursive dominance of American political philosophy (and social-scientific positivism) over political economy (and dialectical thinking) during the twentieth century. As history would have it, the founding parents of modern American political philosophy were John Dewey and William James, rather than Sidney Hook, and the teleological leanings of Dewey and James were progressivist and social Darwinist, rather than dialectical and Marxist.

This historical distance from political economy in North American academia has profound effects throughout popular criticism today. Take, for example, one of the dominant strands of the antiglobalization movement that insists on catching corporate criminals red-handed. Indeed, this is precisely what was presented as the moral force at the beginning of the antiglobalization movement. Naomi Klein's argument in *No Logo*, for instance, was effectively that if young consumers knew from where their clothes came, if they were able to map all of the networks involved in serving their Starbucks coffee, they would be disgusted, their desire for the commodities would whither, they would resist, and they would effectively change the world. Among the three key assumptions of this argument (first, that knowing about how something works will necessarily change how one behaves; second, that changing one's behavior will necessarily lead to systemic social change; and third, that the origin of the corruption is located in the transgressive act itself), the fetishization of the transgressive act concerns us here.

An overemphasis on transgression leads to the following types of questions: "Where is the immediate exploitation of labor?" "Where is the contract being broken?" and "Where is the corrupt official?" These questions must be asked because there is too much injustice that goes on as a result of corruption. When Nike cleans up its act, however, and everything is supposed to be fine (when the contract is obeyed to the letter and the factories are safe) there will still be systemic inequality and environmental degradation due to the larger commodity system. But how are we to articulate what is wrong with it then? To criticize capitalism based on its transgressive acts, therefore, is to forgo an analysis of how capitalism produces inequality even when it works above board.

Does one really need to be a Marxist today to recognize that the system is structured in dominance, structured to produce great wealth for a privileged minority? Does one really need to be a so-called Red Green to recognize that commodity culture is destroying the environment? Does one really need to be trained in dialectical materialism to recognize that war and prisons are not only highly profitable industries but indispensable for the reproduction of capitalism? The answer to all three of these questions is a resounding "No!"—an answer, however, that requires an understanding of crisis that does not get caught up in fires and floods or the shameful deals brokered by Milton Friedman's disciples. As they are for the human subject in terms of psychoanalysis, crises are built right into the system and daily reveal themselves in between the great disasters and capital "C" crises that consume media attention, as well as individual consciousnesses. The lowercase "c" crises, the ones that make up the banality of our everyday lives (the ones that don't even look or feel like crises, but keep the whole system going) consume not only our nonmediatized lives but our unconsciouses. And this is doubly so for disaster.

CRISIS, DISASTER, REVOLUTION

It is at this point, however, that we need to distinguish between disaster and crisis. Disaster is that moment when the sustainable configuration of relations fail, when the relation between one thing and another breaks down.[49] In finance (for a capitalist economy), disaster hits when goods cannot be related to markets, when idle capital and idle labor cannot be connected, or when currency bubbles burst, replacing so much cold cash with so much hot air. In ecology, the disaster of global warming hits when the emission of carbon dioxide no longer relates to the planet's natural capacity to absorb it. For those with HIV or cancer, disaster comes when cells overproduce so that they no longer relate to the logic of the living body, or when one is denied antiretroviral or chemotherapeutic drugs due to the inability to pay for them. In philosophy, disaster is that moment when thinking is cut off from history, while individuals are in psychological disaster when they are no longer able to relate to the world. As for political disaster, it comes with the severed relation between those desiring representation and those authorized to grant it.

One thing we invariably learn when natural disasters strike (such as the Southeast Asian tsunami in 2004, the Hatian earthquake in 2010, or the partial meltdown of the Fukushima Daiichi nuclear reactor in Japan, following the earthquake and tsunami in Miyagi prefecture in 2011) is that such events are not natural, or at least the effects of such events are not natural. Their fallout, quite obviously, is social—products of human choices, political systems, even cultural assumptions. Extending this understanding to the limit, however, effectively evacuates the category of disaster itself. This is because although disaster is contingent (coming "from the stars," as its etymology suggests), its effects are almost always predictable and quite logical. Most people in power knew exactly what would happen if the New Orleans levees broke, just as epidemiologists can predict how many will die of AIDS if left untreated. Those in power simply cross their fingers and hope that such events will not occur. When they do occur and their tragic consequences ensue, calling them disasters is like calling a dying man a hypochondriac.

However much their effects may be completely predictable, the contingency of disaster is what sets it apart from crisis. Unlike a disaster, there is something necessary about a crisis, something true to the larger systemic form. In other words, systems are structured so that crises will occur that strengthen and reproduce the systems themselves. The boom-bust cycle of capitalism is only one of the more obvious examples of this logical necessity. Both contingent disasters and necessary crises, therefore, are linked in the way that their breakdown in relations is built back up again by a different set of relations within the same system.

Revolution, in contrast, is that moment when a new set of relations takes hold within a different system. This crude distinction better explicates the new ubiquity with which disaster and crisis have been invoked over the past twenty years, while revolution has been driven underground, not only rendered unspeakable, but, more important, unthinkable. This trend has everything to do with the political-economic situation of the post–Cold War era, a symptom of our own historical formation, which currently, for good or ill, goes by the name globalization.

Disaster and crisis have always been quick off the lips of those wishing to justify mishap and misfortune. If it were not for that earthquake, the town would not be in such disrepair. If it were not for the crooked officials or crony capitalists, there would be better public transportation, better health care, and more wealth to go around. If it were not for the new terrorists, we would be free from anxiety, sleeping comfortably on cushions bought by the peace dividend. Crisis and disaster are those props pulled out of the bottom of the bag when all other explanations lose operational force or cannot be spoken.

With the end of the Cold War, anomalous and nonsystemic disaster and crisis (that is, events from the outside, like a meteor or a madman) have been even more likely to be employed to explain inequality and injustice. During the Cold War, for example, to speak the language of disaster and crisis was at once to speak the language of revolution: the discourse could easily slip into revolution. Disaster and crisis were truly dangerous. With "mutually assured destruction" the watchwords of the day, one crisis could accumulate into so many crises until the quantitative curved into the qualitative and the whole system was in tatters. We only need to think about the Cuban missile crisis or the oil crises of the 1970s to remember what a cat's step away crisis and disaster were from revolution. But with the transformed geopolitical situation following the Cold War, in which the United States was left as the sole superpower and the "end of ideology" became the ruling ideology, it *seemed* riskless (not to mention utterly gratuitous) to call upon crisis and disaster.

Following the Cold War, crisis and disaster were as far apart from revolution as heaven from earth. What needs to be considered in the current post post–Cold War moment is whether or not this is still the case. Is something changing so that crisis and disaster are becoming dangerous again, no longer the trump cards of those in power? Is something changing so that revolutionary discourse is creeping back into everyday consciousness, into the way we understand not only radical social change but the more banal ways we understand ourselves and think about the future?

Something is changing, but this can only be understood when we understand crisis and disaster as nonspectacular and in everyday terms. This is the present challenge for counterglobalization movements: to bracket the more spectacular political scandals, corruptions, and disasters and turn to everyday economic activities (and the crises inherent in them) in order to understand larger formal problems. Instead of treating the political spectacles as chronic (as problems to manage), the economic minutiae should be treated as acute (as something to be radically changed). Which returns us to revolution.

Tracking how revolution is understood today is essential to properly mapping contemporary intellectual thought. Among the intellectual left, key differences can be discerned by focusing on what a thinker makes of revolution today, how such a theorization relates to past theories of revolution (and actual revolutions), and how such an understanding shapes the importance and privilege granted to particular disciplines (such as economics, politics, philosophy, and culture). In what follows, I will briefly track key similarities and differences on these three points by referring to some of the more significant and influential theorists of revolution today, Michael Hardt and Antonio Negri, Alain Badiou, Slavoj Žižek, and Karatani Kojin. My main point is to mobilize their work on revolution in order to speculate on what it might open up for the analysis of illness and, more importantly, to further develop the comparison I am making between revolution in politics and cure in medicine.

When writing about a method for generating institutional reforms in *Multitude*, Hardt and Negri insist:

> There is no conflict here between reform and revolution. We say this not because we think that reform and revolution are the same thing, but that in today's conditions they cannot be separated. Today the historical processes of transformation are so radical that even reformist proposals can lead to revolutionary change. And when democratic reforms of the global system prove to be incapable of providing the bases of a real democracy, they demonstrate ever more forcefully that a revolutionary change is needed and make it ever more possible. It is useless to rack our brains over whether a proposal is reformist or revolutionary; what matters is that it enters into the constituent process.[50]

By "constituent process" Hardt and Negri mean that there is a single creative force and that emancipation comes out of the dynamics of contemporary capitalism itself. Anyone who believes that such change can come from outside capitalism (that change is not immanent in the system itself) is just another schoolboy desiring a new master. For Hardt and Negri, holding the binary of reform and revolution (and to privilege revolution, as the old left insisted on doing) is therefore tantamount to a form of religious belief. At the same time, Hardt and Negri retain a fundamental critique of capitalism in which the economic dimension and the historical specificity of global capitalism (in terms of new forms of labor exploitation and geopolitical relations) cannot be underestimated.

Alain Badiou sees Hardt and Negri's view as too systemic, something akin to a revolutionary determinism (not unlike the First International progressivism that viewed the contradictions of capitalism as necessarily leading to communism). For Badiou, this overlooks the necessity of a revolutionary event that cannot be predicted, comes from nowhere in particular, and inhabits a singular time outside of the reigning temporality of any dominant situation. As Badiou puts it: the event is "a totally chance, incalculable, disconnected supplement to the situation." The idea of the revolutionary event, therefore, distinguishes Badiou from Hardt and Negri and their uncompromising immanence. But, Badiou would deny that he is merely waiting for the magic transcendence of the revolution. "Now I am absolutely an immanentist," Badiou states, "I am convinced that if there is truth, it isn't something transcendent, it's in the situation—but I am nevertheless led to the conclusion that the situation, as such, is without truth. This antinomy must be resolved. That's where I turn to the category of the 'event,' which pushes the system in another direction."[51]

Slavoj Žižek differentiates himself from both Negri and Hardt and Badiou. He agrees with Badiou about the revolutionary event that "cannot be reduced to its causes or conditions" and that such an event holds open the only way that substantive change can come into the world.[52] Žižek also insists on the possibility of rupture, but one that is not merely wishful thinking, as Hardt and Negri might charge. Žižek, nevertheless, comes closer to Hardt and Negri in how he privileges the economic, or at least in his respect for the horizon of capital as that space in which any radical change must be theorized and enacted—not as mere background but as "the secret point of reference and structuring principle of political struggles."[53] Unlike Badiou, who rejects a generalized anticapitalism as a structuring principle (believing that any intervention on the level of economics is ultimately a throwback to an earlier moment of revolutionary politics that has not been appropriate since Lenin) and who advocates a pure politics of prescription that privileges single issues and relentlessly pursues them without getting side-tracked by too many generalized political pronouncements, Žižek does not give up on the party or on older revolutionary models translated into the present. Žižek likes to remember Beckett's saying, try again, fail again, fail better.

Karatani believes, like Žižek and contra Badiou, that the economic presents the most appropriate space for resistance. For Karatani, it is in the circulation process itself where new associations, organized around an alternative paradigm of exchange, struggle both *within* and *without* the capitalist mode of production: "The struggle immanent in and the one exscendent to the capitalist mode of production/consumption are combined only in the circulation process, the topos of consumers=workers."[54] This also reveals a key difference with Hardt and Negri, at least in terms of how Karatani argues that the exscendent of capitalism (a blend of the words "exit" and "transcendent"), what Hardt and Negri would challenge, is key to making radical change. Moreover, Karatani sees this parallaxical point of leverage between production and consumption as already presented in Marx's *Capital* and as most promisingly practiced in the contemporary movements following an associationist model, an economic alternative of networks and services that challenges what Karatani calls the interlocked trinity of capital, nation, and state.[55] Karatani develops this analysis in his book *Sekai-shi no kozo* (2010), in which he recognizes that it was Hegel who, in *The Philosophy of Right*, had already theorized the importance of exchange, but (and unlike Marx) did not combine this with a strong critique of the state.[56] By advocating for new models of exchange, for what he calls exscendent counter-acts (based on a local exchange trading system model), together with a trenchant critique of the state within the context of global capitalism, Karatani argues for an alternative economy and, ultimately, for an actually existing outside of capitalism.

All of these thinkers struggle with how to explain change, how to account (if not open up the very possibility) for something new to come into the world. And they all are continually building on their work and retheorizing change in relation to the changing historical situation. My point is not to reduce their theorizations of revolution, but to speculate on how they might translate into the medical realm. For example, how might these different ideas of immanence and transcendence, of revolutionary events and strategic withdrawals, relate to the role of cure and management within contemporary medical discourse and practice? And, more important, what is at stake in this relation between the medical and political realms?

In *Transcritique* Karatani writes about cancer as follows:

Karl Polanyi likened capitalism (the market economy) to cancer. Coming into existence in the interstice between agrarian communities and feudal states, capitalism invaded the internal cells and transformed their predispositions according to its own physiology. If so, the transnational network of workers qua consumers and consumers qua workers is a culture of anticancer cells, as it were. In order to eliminate capital, it is imperative to eliminate the conditions by which it was produced in the first place. The counteractions against capitalism from *within* and *without*, having their base in the circulation front, are totally legal and nonviolent; none of the three can interrupt them.[57]

For Karatani, by extension, to eliminate cancer it is imperative to eliminate the conditions by which it was produced in the first place, involving everything from ecological practices to the very discourses of cancer that are indispensable in making cancer so deadly and feared. But such an imperative requires retaining the category of cure, and even retaining the cure-management binary. Hardt and Negri, in contrast, would see little need for retaining the binary, because in today's medical environment (with such advanced technologies and drugs) the categories of cure and management cannot be separated. They would probably be more sympathetic to new designations, such as "functional cure" or "sustained management," in which the management of symptoms is so successful that the disease itself, if not the very logic of illness, is radically transformed. Žižek, in his inimitably dialectical way, would probably turn everything around by arguing for a medical goal that cures functionality—arguing that it is cancer itself that produces that illness called functional health. He might focus on how this disease can turn back on and transform the very real effects of cancer. To eliminate cancer, in other words, one would have to eliminate not only the conditions by which it was produced (as Karatani argues), but its supplementary twin—non-cancer.

This takes us to Badiou and his politics of prescription. A prescriptive politics (not unlike prescriptive medicine) is targeted and specific to a situation. It is the application of an axiom to the here and now. One begins with, say, equality and then engages an issue such as health care based on this axiom. Every justification for why some receive life-saving medications and others do not falls short: either one believes in equality (evidenced by how one follows through with the prescription) or one does not believe in equality. There is no hiding behind deferral, long-term goals, or other apologies.

Like prescriptive medicine, this prescriptive politics came into being with the end of the Cold War. The revolutionary discourses associated with the Cold War (primarily based on political ideologies of democracy and freedom, either of the capitalist or socialist kind) transformed during the post–Cold War era into either antirevolutionary discourses (based on political ideologies trumpeting the final victory of global capitalism) or into the postrevolutionary kind in which single issues are targeted, engaged, and managed like so many mutated chromosomes. Indeed, the global market has shifted the geopolitical system so that national liberation and nation-state-centered revolutionary possibilities have shifted as well. I do not have trouble rethinking revolution in terms of this shift. And I do appreciate the more progressive aspects of a prescriptive politics, not to mention the life-saving drugs of prescriptive medicine. But I want to emphasize the importance of retaining the categories of revolution and cure today. Revolution and cure are not only about politics and medicine, but profoundly shape consciousness itself, so that one's abstract ideas about these categories significantly inform ways of thinking and the imagination in general, especially about nonrevolutionary life, no matter how hackneyed and everyday.

It was Georg Lukács who made this point when arguing that the point of view of totality ("the all-pervasive supremacy of the whole over the parts") is the bearer of the principle of revolution. Marx made little analytical advance over the classical political economists, at least in terms of the discrete categories of the discipline. Rather, it was the way Marx put the categories in relation to each other and in relation to time and history itself (the way he was able to abstract the system as such) that made his work revolutionary. What is revolutionarily possible on the ground and what is revolutionarily possible in consciousness do not necessarily move at the same speed. My point is that when retaining the primacy of revolutionary thought today, we return to time its radical historicity, in this case the historicity of globalization itself. That is to say, we retain the imaginative space to consider the end of globalization, a social dreaming that has very real consequences for how we can think, and how we can act, in the present. It is for this reason that I am also skeptical about giving up on cure altogether. By retaining cure we retain the historicity of illness—the idea that illnesses (like capitalism) have their own histories and are inextricably tied to history. HIV is as much a symptom of globalization as it is of immune deficiency. It is true that cure was never a stable category even at the best of times, but when we dismiss it altogether we risk losing the capacity to think radical difference and to imagine an alternative to the here and now—a failure of the imagination that is exacerbated by the anxieties associated with dominant ideologies of life and death. Today, there is a discursive struggle within the political and medical realms—the stakes of which come into full relief when we place these realms side by side and attend to their formal similarities.

Revolution and cure are not just about politics and medicine, but shape the more general ways we imagine, understand, and act in the world. In the early 1990s, Jameson commented on the lack of a revolutionary imagination in his now famous statement that it is easier to imagine the end of the world than the end of capitalism.[58] The force (and humor) of this thought momentarily revealed the depoliticized nature of late capitalist societies, always lulled by the latest blockbuster dystopia as a way to stay asleep to the actually existing possibilities of radical change. But today the dialectic shock of this statement seems to have worn off. Yes, it is easier to imagine the end of the world (or at least the species) than the end of capitalism. Why has Jameson's joke stopped being funny and become for many scientists a statement of fact? Because the end of the world *is* a more likely scenario than the end of capitalism. Sadly, capitalism might very well be the last mode of production during human history. What might such a bleak prospect teach us about the possibility of radical change?[59] Or, we can go one step further and ask what radical alternatives might actually be enabled by such a bleak prospect—as if in some grand dialectical reversal in which the only way to make it out alive is to genuinely submit to one's own death.

To develop this let's turn to the financial crisis of 2008, probably the only moment since the crash of 1929 when imagining the end of capitalism seemed most possible. In fact, it really was the end of capitalism, at least the end of a certain imagination of capitalism. People used to believe that capitalism was primarily a political force, that it enabled democracy, equality, liberty, justice, individualism, and other key political categories most powerfully mobilized during nineteenth-century imperialism and again during the Cold War. But what has changed over the past twenty years or so is the recognition that capitalism is what it has always been, primarily an economic force. It is a force that is foremost about commodity production, expansion, and the necessity of profit creation, as well as the need for ever new modes of accumulation that have reached schemes of financial speculation of the highest order. We might even want to date this end of political capitalism to July 18, 2009, when Goldman Sachs announced $3.44 billion in quarterly profits (33 percent over the first quarter) just nine months after receiving $10 billion worth of the Troubled Asset Relief Program loans provided by U.S. taxpayers to bail it out. Goldman Sachs benefited further when the Securities and Exchange Commission barred investors from betting against Goldman's shares by selling them short, as well as when Goldman was given permission at the height of the crisis to convert from an investment firm to a national bank, making it easier to access federal financing in the event it came under greater financial pressure.[60] Goldman's announcement came at a time when unemployment in the United States was nearing 10 percent, when home foreclosures were still rampant, and when food stamp enrollment had reached a new record at 34 million.[61] Since the average benefit per person for the food stamp program, renamed the Supplemental Nutrition Assistance Program in October 2008, was $133.65 per month, this means that Goldman Sachs's second-quarter profits alone could roughly fund the entire program. How could Goldman report the biggest quarterly profit of its 140-year history and set aside $11.4 billion for employee compensation when the United States and the rest of the world were reportedly still in the midst of the greatest crisis capitalism had faced in its over 400-year history?[62]

Only two weeks before Goldman's report, a young *Rolling Stone* reporter named Matt Taibbi published "The Great American Bubble Machine," in which he called out Goldman Sachs on its suspicious knack for cashing in on capitalist crises. Taibbi called to the table all those suspected of what appeared to be a return to business as usual on Wall Street.[63] Taibbi was not alone with this critique. The *Wall Street Journal* declaimed, "We like profits as much as the next capitalist. But when those profits are supported by government guarantees or insured deposits, taxpayers have a special interest in how the companies conduct their business." The *Economist* magazine contended, "Such largesse is cheeky at best, distasteful at worst." Fox News' Glenn Beck held forth on "Why Goldman Sachs Is the Evil Empire!" Paul Krugman of the *New York Times* noted, "The huge bonuses

Goldman will soon hand out show that financial-industry highfliers are still operating under a system of heads they win, tails other people lose."[64] Very few people were coming to Goldman's defense, except, of course, for Goldman Sachs itself. A Goldman spokesperson told one reporter, "Taibbi's article is a compilation of just about every conspiracy theory ever dreamed up about Goldman Sachs, but what real substance is there to support the theories?"[65]

To the charge that Taibbi's excoriation of Goldman was made up of garden-variety conspiracy theories, former attorney general and governor of New York Eliot Spitzer (called the sheriff of Wall Street before stepping down due to scandal) responded with the following statement: "Because it is a conspiracy does not mean it is wrong."[66] This is a brilliant statement, despite being entirely wrong. To explain why the statement is wrong, we have to explore the concept of conspiracy itself and the logic of conspiracy within capitalism; but in order to do so we must first track the dominant ways by which critics have accounted for Goldman Sachs's success.

The question boiled down to how Goldman Sachs made so much money when other companies were taking major losses or had already folded. Those answering were split: on one side those defending Goldman argued simply that the employees of Goldman Sachs were smarter than the rest, that they were the best and the brightest and were able to make better decisions than their competitors, namely, their main rivals in Bear Stearns and Lehman Brothers (who both went bankrupt). The other side explained that Goldman succeeded primarily because it stacked government with its former employees, who then went on to shape policy and make decisions that benefited Goldman and imperiled its competitors. The prime example was Henry Paulson, secretary of the treasury under George W. Bush. Paulson was the main architect and defender of the bailout, and he was the former CEO of Goldman Sachs. The fact that Paulson agreed to bail out AIG for $85 billion (which promptly went on to pay off a $12.9 billion debt to Goldman) and let Bear Stearns and Lehman Brothers fold, as well as the fact that Paulson was on the phone with Goldman CEO Lloyd Blankfein two dozen times during the week of the AIG bailout (and two times before receiving a waiver that allowed Paulson to deal directly with Goldman due to conflict of interest concerns) only gives credence to this position.[67] But where the "smartest guy in the room" position and "the most corrupt guy in the room" position come together is around the importance granted to Goldman executives (either as brilliant or sinister) and the insistence (however unconscious) on analyzing the capitalist system only in moralizing terms. Again this is a hangover from the nineteenth century and the twentieth, in which capitalism had been understood primarily as a political, rather than an economic, force and in which the arguments for capitalism and against socialism were political and moral arguments, rather than economic ones.

Returning to Spitzer's point that conspiracies are not always wrong, we can now argue the opposite: conspiracies are always wrong. For a conspiracy to work there must be something that cannot be analyzed, that cannot be figured out. Conspiracy theorists are driven by instincts and hunches, and despite the fact that those hunches are corroborated by smoking guns and whistleblowers, they are never totally confident in their accusations. The conspiracy text (which is the form of representing the conspiracy theory) may, according to Jameson, "be taken to constitute an unconscious, collective effort at trying to figure out where we are and what landscapes and forces confront us in a late twentieth century whose abominations are heightened by their concealment and their bureaucratic impersonality. Conspiracy film takes a wild stab at the heart of all that, in a situation in which it is the intent and the gesture that counts." Jameson continues by arguing, "Nothing is gained by having been persuaded of the definitive verisimilitude of this or that conspiratorial hypothesis: but in the intent to hypothesize, in the desire called cognitive mapping—therein lies the beginning of wisdom."[68] Conspiracy theories are always wrong because they are less about the conspiracy itself than about a desire to map an unmappable system. And this brings us to what distinguishes our present moment from the emerging postmodern moment that Jameson refers to in his analysis of conspiracy films from the 1970s and 1980s.

What marked the various conspiracy texts that Jameson analyzed, such as David Cronenberg's *Videodrome*, Alan Pakula's *The Parallax View*, and Francis Ford Coppola's *The Conversation*, was the new global integration of power and connections that exceeded the former modern configuration dominated by the nation-state. Up until the 1970s, cognitively mapping the system required tracking the class relations that structured a respective nation (relations that mediated everything, from urban design policy to the narrative choices of television programming to the racialized and gendered division of wealth) and how those national relations shaped and were shaped by other nations. But what the conspiracy films of the 1970s and 1980s so powerfully revealed—in the very way they failed to represent the whole conspiracy—is that a new configuration was at work, one whose disparate connections (those that exceed not only individual nations but the very nation-state system itself) cannot be drawn together in any coherent way. Today, however, things have shifted once again so that it is now possible to draw a relatively coherent map of the global system, which also means that newly unmappable processes have emerged. This historical turn, moreover, transforms the possibilities and limitations of the conspiracy text itself. Take, for example, *The International*, a contemporary conspiracy film made in 2009 by German filmmaker Tom Tykwer.

The International is about the corruption and global reach of the International Bank of Business and Credit (IBBC), registered in Luxembourg and involved in everything from insider trading to the financing of terrorism. The fictional IBBC is based on the Bank of Credit and Commerce International (BCCI), which was founded in 1972 by Agha Hasan Abedi and went defunct in 1991. As one report to the U.S. Senate charges, "BCCI's criminality included fraud by BCCI and BCCI customers involving billions of dollars; money laundering in Europe, Africa, Asia, and the Americas; BCCI's bribery of officials in most of those locations; support of terrorism, arms trafficking, and the sale of nuclear technologies; management of prostitution; the commission and facilitation of income tax evasion, smuggling, and illegal immigration; illicit purchases of banks and real estate; and a panoply of financial crimes limited only by the imagination of its officers and customers."[69] After reading this litany, it is hard not to recall Bertolt Brecht's only half-mocking question from *Happy End*: What's the difference between robbing a bank and founding one?[70]

Despite Brecht's insight about the logic of capitalism and the dialectics of crime, the revelation of the corruption behind the BCCI in the 1980s and early 1990s was truly stunning and symptomatic not only of the new global reach of financial criminality, but also of the Reagan-Thatcher neoliberal project of near total deregulation. The creators of *The International*, however, chose to contemporize the events and push everything into the post-Madoff climate of 2009. It is for this reason that the film fails to deliver the same tight-fisted intensity and this-can't-be-happening incredulity that the best films in the genre do. The point is that nobody seems shocked that such a bank might exist today. Such banks do exist, as the crisis of 2008 revealed by exposing how exotic mortgages were carved up and resold so many times that an investment firm in Pakistan can own suburban Californian homes. And a small Californian bank is just as likely to have a stake in Taliban financial dealings. The revelation of *The International*, therefore, cannot come off as anything but quaint. The moment one can name the conspiracy and clearly represent it is the very moment that the conspiracy no longer exists. Or it has turned into some other unmappable conjuncture that can only be seen by those with the desire to do so. Conspiracy theories are always wrong.

But the desire to map out a conspiracy is always right, right in the way that it expresses a wish to break out of an ever-tightening discursive space. Thus, the lesson to be learned about the current moment comes less by way of Goldman Sachs and more by the debate over health care in the United States. The most interesting effect of the Obama health care plan, debated in the summer of 2009, was how far away the popular debate over the proposal was from the actual plan itself. Angry audience members shouted down politicians of all stripes during raucous town hall meetings. From the threatening dystopia of "death panels" to the cartoonish nightmare of "socialism," the rhetoric of the shouting match completely occupied any space for critical debate.

One way to make sense of the situation is to recognize that the intense reaction to Obama's health care proposals was not about health care at all, but about deeper ideological pressures that generated such angry and self-righteous affect—affect that is supercharged and circulates at remarkable speed due to the twenty-first-century mediascape. Some have argued that the resentment was more about larger-scale social transformations and the loss of entitlement that many older white voters felt. When comparing the relative lack of anger toward the financial bailouts to the shrill response to health care reform, Jodi Dean puts it this way:

> The racial politics at work might account in part for the failed outrage around the bailouts—bailouts meant that white people were still ahead, were winning, were on top. Same with outrage fail over bonuses and the general Wall Street pillage: done by white people so still part of white hegemony (class struggle has always been ruptured by racism in the US). But, with health care, the basic affective dimension is that the

white winners are having to pay for black and immigrant losers. This means that the worried whites are dealing with the barely repressed knowledge of their economic misery—their tax dollars paying out banks —and with seemingly undeserving black people, who are going to get something for nothing.[71]

The racial component is important and the fear of scarce and ever-diminishing public resources is real. But we need to push this even further and argue that the failed outrage over the economic bailouts is not only due to white resentment and fear of a quickly weakening majority, but to a certain ideological understanding of capitalist economics. The taboo of having any real discussion of capitalism is so entrenched in American society that every other issue, issues as varied as health care, education, and the military budget, can never be properly debated or understood.

When the protesters question health care reform and worry about being on the wrong end of rationed care or a government-run option that might deprive them of certain procedures, they invoke the political categories of liberty and freedom. And yet when they take to task a privately owned insurance company for rejecting their medical claims, they invoke the economic categories of profit and efficiency or greed, as they did when United Health Care CEO Stephen Hemsley received almost $750 million in stock options while many of his financially strapped customers were ter-rified that they would not be able to afford life-saving drugs after their claims were denied. But now we have returned to a moral register, in which a more objective analysis of how capitalism enables—and relies upon— such astonishing divisions of wealth between Hemsley and a minimum-wage worker is left undeveloped.

Why is it that capitalism cannot be seen for what it is? Not as a good or bad system, but as a system with a certain logic built right into its most fundamental part—not the cold or warm hearts of its human subjects but the commodities that must continually be bought and sold. Is not the conspiracy today the very fact that there is no conspiracy today, but a system that works precisely as it is intended to work? This is why conspir-acy films today do not quite work, at least not those concerned with the "boardroom conspiracies" of greedy CEOs. Mapping the system is a banal task; the map, we might say, is already published and explicit, so that we are left with the question, "Given what is permissible and in plain sight, what would CEOs need to conspire about?" And thus it is finally the conspiracy theorist himself who becomes the conspirator (witness Steven Soderbergh's *The Informant*).

This brings us back to the chronic, for it is this new configuration that today closes access to the required analysis of capitalism. The chronic keeps the conspiracy intact, as if somehow, because we now *know* how the profit motive and commodity production work to regulate our economy, we can learn to live with them, to manage them. Once the malady is identified and understood, we can adjust ourselves to its presence. In place of this chronic logic, a new appreciation of the terminal is required, which leads to the belief that an open acceptance that the end of the world is near can be the most radically progressive position to take (and life to live), one that can ironically provide us with the best chance of getting out of this mess alive.

How might this understanding of the shortened future shape our relationship to the present? This is not an argument for some kind of eco-apocalypticism that celebrates an end or resigns itself to the status quo, but for a new appreciation of the meantime, of living in the midst of crisis, a mode of being that does not give up on change (on cure), but at the same time is not paralyzed by it. There is a lot to learn from certain medical models, such as those managing terminal illnesses and the fields of palliative medicine and psychoanalysis that have already been discussed. There is also much to learn from certain political events, events demonstrating the discursive limits and possibilities of reform and revolution. But instead of returning there, let's move to the cultural.

The Cultural

To begin examining this issue of culture and the chronic, a return to the theme of illness is in order. At first it seems that a cultural study of illness would entail carving out the genre of illness narratives (both fiction and non-fiction) and arguing for how representations of illness reproduce or challenge dominant ideologies of health care. Instead of studying narratives about illness, however, I will come at this from the other direction and look at the way dominant discourses of illness (namely, the new chronic mode itself) inform narratives of all kinds. In other words, when we step back from the specificities of contemporary cultural works (from the different aesthetic practices that distinguish literature from music, film from dance, high art from popular art) we can detect a more general trend today shared by all of these forms, one that relates to the organization and experience of time. In particular, I am referring to the way the computer and digitization in general brings with them a certain time that is at once instantaneous and infinitely deferred, singular and endlessly hyperlinked, terminal and chronic, but a time becoming more and more weighted toward the chronic at the expense of the terminal, not unlike the tilt toward the chronic in contemporary medical and political discourses and practices.

For example, take the boom in what is called reality narratives (real-time reportage, amateur videos, surveillance, live web cams, and, perhaps the pure form of these narratives, images of actual death). This is a form in which a crisis is not just being recorded but simultaneously produced in order to service a market of viewers who desire to experience these events (a market that is mediated less by the television or the film screen and more by the computer monitor and cell phone display). The allure of reality culture is that events occur unpredictably, despite the fact that reality culture attempts to commodify and reproduce these seemingly irreproducible events. By manufacturing crises, reality culture effectively attempts to preempt them—for crises are precisely those events that cannot be contained, reproduced, or commodified. The crisis culture that streams directly to our computers (bombs exploding, earthquakes destroying, individuals injuring) does not represent a short-term danger, but its opposite, a long-term safety insofar as it attempts to arrest the future by the very way it shores up our deepest fantasies and fears. We can prepare for crisis, we can stage it, reenact it, even practice it, but when the airplane is going down we can never be sure if we will help others out the door or be pinned to our seats in uncontrollable fear. And, oddly enough, it is precisely this radical contingency that is most utopian and most at risk of being buried by chronic culture.

This sea change, brought about by contemporary media practices saturated by an ever consolidating and hypercommodified global mediasphere that now informs how we experience even nondigital and predigital culture, is organized around a dominant operation of time that works (on one level) to annihilate the future and, ultimately, eradicate our most intimate relation to death—an intimacy that when snuffed out leads to aggression and violence of heartbreaking proportions (violence toward nature, toward others, toward ourselves). But this operation of time also, as exemplified by reality culture, generates a radical dimension: the desire (however unconscious) of its viewers for the everyday to erupt, to explode the chronic temporality that reality culture is instrumental in producing in the first place.

This double dimension is not unlike the radical and reactionary dimensions that we saw in prescriptive medicine and prescriptive politics. Although the dominant effect of prescriptive medicine, for example, is to control medical practice by pharmaceutical corporations and to consolidate the chronic, the targeted drugs themselves also contain the possibility to escape the chronic by producing a medical event that fundamentally changes the nature of a disease itself (as has happened for some cancers and HIV). Likewise, a reactionary prescriptive politics is one that is occupied by a single issue in a way that effectively strengthens the larger system that produces and solidifies that very issue itself. Badiou's more radical prescriptive politics, on the contrary, pursues a single issue with the larger aim of provoking an event so substantial that the very nature of the issue at stake, and everything else, is utterly reconfigured.

As it is in the medical and political realms, a more radical prescriptive culture is emerging—one that is targeted and management based. It intervenes in local situations and resists any attempt at generalization. But I am not referring now to reality culture, but to a more experimental and noncorporate culture that is best represented by a certain type of "network art," an art that is provisional, process guided, collaborative, short-lived, irreproducible, and uncommodifiable, a new type of media art that requires users to participate in the production of one-of-a-kind artifacts, thus undermining repetition and reproducibility, if not the market itself. Any cultural intervention that is not singular and immanent (the two dominant theoretical watchwords of our present moment) is thought to be a betrayal of art and a reactionary return to the worst kind of instrumental, state-controlled culture. It is argued, therefore, that postinstrumental, liberated culture must be targeted with no claim to anything other than its specific intervention—however much it generates a certain translocal reciprocity, multiplying and intensifying connections like so many Deleuzian rhizomes. At its best, this type of prescriptive culture is truly radical, as Peter Mörtenböck and Helge Mooshammer argue in *Networked Cultures*, a book that documents self-organized, transversal, and autopoietic art projects: "Such a 'disciplineless' praxis of unsolicited intervention in spatial contexts renders legible the dysfunctional rules of planned spatial and cultural containment and creates an avenue for generating new forms of circulation amidst the political efforts to conceal this failure."[72] Indeed, the real-time, live aspect to this culture resists the more reactionary dimension of reality culture I described earlier. The logic of capital, however, cannot be so easily circumvented, as there are ingenious ways the market appropriates such anticommercial forms. If capital can market crisis and disaster (if not death itself) then surely it can recuperate its harshest cultural critique.

Many of these new projects are breathtaking and admirably require us to reconsider the classic relation between aesthetics and politics. The *Lost Highway Expedition*, an aesthetic and social experiment carried out in 2006 involving a collective journey through a Balkan highway, a journey started in 1948 but never completed, is just such a project. The *Place of Solitude*, a collaboratively developed space on the outskirts of Paris where new immigrants can escape the city as well as the confinement of their own homes and communities, also redefines the relation between aesthetics and politics. I would even argue that this type of impressive art practice has reached its historical limit; it goes as far as any committed, politically significant and aesthetically inventive art can go today. But this is not far enough. As I did in regard to the medical and the political, I wonder what is at stake when we give up on the classic category of revolution and invest all our energy and resources in local, nontotalizing cultural targets. Could there be in the near future a "planned culture" or a "cultural revolution" that overcomes these current limits and radicalizes what now seems to be the telos of radical art? If so, and even though we might not be able to currently imagine what such a revolutionary culture or cultural revolution might look like, we can be sure that it will look nothing like what we already know—no overly zealous Red Guards smashing Ming vases, no genius artists smashing aesthetic conventions. Finally, if such a revolution does come into being it will necessarily have to challenge the new chronic mode. It will shatter the secret identity that connects the temporality of an individual illness to the temporality of a capitalist investment and the temporality of a cultural narrative to the temporality of a political action.

To analyze the various challenges to chronic culture today, I will explore the relation between revolution and culture in three ways. First, I will engage the revolutionary study of culture by focusing on transformations within academic humanities departments. Second, I will explore revolutionary culture by looking at the transformations in the genre of amnesia films, beginning in Hollywood and then moving to Japan, before arriving in Britain with an analysis of Manu Luksch's *Faceless*, a film composed solely of British surveillance camera footage that narrates a science-fiction story about a memoryless society dominated by a chronic temporality. Third, I will engage the problem of cultural revolution today by analyzing the manifesto form and how certain aesthetic-political projects today relate to past avant-garde manifestos and to the current moment in which cultural manifestos themselves seem to represent not only a throwback to an earlier moment, but a betrayal of the last remaining free zones from capitalist instrumentality—experimental art and anarchic politics. But the manifesto today does not need to be such a betrayal, as Luksch demonstrates with the manifesto she includes with *Faceless*. The cultural manifesto is most radical today precisely because it seems to betray everything we have learned from the twentieth century and pushes beyond network art to reintegrate aesthetics and politics, theory and practice, the chronic and the terminal, all while retaining the integrity and unevenness of the two categories in potentially explosive ways.

VULGAR CULTURE, AMNESIA, THE CLINICAL

The old insult that a vulgar (or revolutionary) cultural analysis has been performed was usually waged against critics who "shot too soon" from the aesthetic to the sociopolitical, critics who ran roughshod over the formal aspects of the text and clumsily plugged its content into a larger historical preoccupation. The text would invariably be considered a symptom of something else, but without much attention granted to the specificity of the symptom itself. This quick and many times liberating move to the social text came in response to a regime of close reading in North America (hailed by the New Critics) that insisted on the text's autonomy and resisted any emphasis on the text's outside. From the 1970s onward, literature departments were training students and redefining literary analysis in the context of this transition. And one outcome of this shift has been a generation of literary scholars (many of whom associate in one form or another with cultural studies) who prefer not to (or are unable to) perform close formal analyses of a given text. Rather than produce a false binary of close reading and a sociopolitical analysis, my purpose is to argue for their complementarity, especially today in an academic climate in which it is de rigueur to defend every critical "intervention" as political, if not radical, thus putting into question (once again) the very relation between politics and culture.

To be revolutionary on the level of a cultural analysis today is to retain the integrity of the levels, the cultural and the political-economic. What I mean by this counterintuitive claim is that rather than reduce the one to the other so that we end up with the true—but empty—claim that all culture is political or that all politics is cultural, we strategically separate out the levels and develop each with a discrete rigor *before* jumping the space between them. Rather than a tenuous interdisciplinary, this would call for a more solid multidisciplinarity, one that would require the humanist to be fully at home in the social sciences, just at it would require the social scientist to employ the aesthetic as something more than just another expedient example. This does not have to reinforce the dominance of the traditional disciplines; rather, it would challenge them from within with a larger project in mind This approach argues that shared forms circulate at different scales (local, national, and global) and at different levels (in the psychological, social, medical, political, and cultural). These shared forms sometimes align to reveal not only the logic of a given historical structure but this very structure's inevitable and always possible unstructurings.

Take just one example: in the 1970s and 1980s the literary critic Masao Miyoshi began to emphasize difference as a strategic political act when engaging in cross-cultural analyses. By arguing that the Japanese prose narrative should not be so quickly identified with the Western novel, Miyoshi exposed how appeals to universality were nothing more than thinly veiled rationales for domination. To delink and emphasize difference within area and literary studies at the time was one way to expose the violence implicit in modernization theory. This move was also a way to challenge the dominant temporality of this model of literary scholarship, which was chronic insofar as any future non-Western cultural product would always be judged in terms of the present aesthetic criteria, criteria that were necessarily produced by metropolitan critics and thus placed metropolitan authors at the top of the hierarchy.

By the 1990s, however, the emphasis on difference had lost its progressive edge for Miyoshi and turned into a reactionary form of humanities-based criticism, one that argues for uniqueness and radical difference in a way that meets up not only with neoliberal versions of multiculturalism but with the most insidious neonativist discourses. The problem now becomes how to think similarity and continuity (how to think totality) without forgetting the sordid history of so much universalist desire.[73] And how does one think totality as a challenge to chronic time, as a formal emphasis of scholarship that opens up the future of literary and cultural production to a radically different totality? How to return to totality an acute temporality? Miyoshi's almost unthinkable solution is a uniquely transdisciplinary and global environmental protection studies that would radically reconfigure the various academic disciplines and merge the humanities and the social and hard sciences. Literature, history, political science, and other disciplines would not simply tinker around their canonical edges (bringing an already established rigor to ecological problems), but would allow the perspective of ecological protection to transform the traditional disciplines themselves.

For example, those in economics would be required to integrate into their discipline a rigorous reflection on, say, geology and philosophy, just as those in literature would have to study the fundamentals of economics as something more than just another object for a sophisticated discourse analysis. This is what it means to retain the integrity of the disciplines and develop each with a discrete rigor before jumping the space between them —but the rigor is doubly required by those scholars outside of each respective discipline. At the heart of Miyoshi's proposal is a hard look at the possibility of human extinction and what other life forms may exist after the end of humans and before the end of the planet. Miyoshi writes, "For now, a new kind of environmental studies will need to decide whether human extinction is worth thinking about. . . . At this time of very little hope all around, we can at least look forward to ongoing life on earth. And as long as we entertain any hope, we will manage to find the courage to keep trying."[74] In this case, genuinely contemplating and studying our potential termination (the death of the human species) is a way to return to scholarship and existential being a project that is alive and critically significant.

This particular relation to time, to the meantime, and to the future is not coincidental, but a symptom of a larger totality that (however inaccessible and open) we ignore at our own peril. Today, to be revolutionary might very well be to reengage this embarrassing problem of the totality, a problem that (for many good reasons) was closeted during the past few decades of cultural and political theory, but now reopens to offer new critical possibilities at a moment of intellectual and political-economic transformation.

As for revolutionary culture today, we cannot analyze its possibilities without analyzing how culture functions within the space of what is called globalization. Only a few years ago, an accessible language did not exist to properly articulate what was going on in the world. No intellectual mouse-trap could catch the palimpsestic disorder of globalization's effects and the subjective experience of these effects. In other words, we were stuck with older forms to envision the new, just as we are stuck with computer key-boards whose arrangement inefficiently imitates that of the manual type-writer. Saddled less with a lack of imagination than a structural block, we were simply unable to imagine and represent the transformed dynamics of the contemporary world-system.

This epistemological and representational limit brought many to the aesthetic, and to the visual aesthetic in particular. For it is the aesthetic that can flash, that can cross over this limit, that can shake this absent, unrepresentable totality into a crystalline moment of dazzling sense.[75] In other words, one can read from contemporary visual culture the dreams and nightmares, the limits and possibilities, of the moment. It was this recognition that led many scholars to science fiction, and to the cyborg narrative in particular, as a rich form to conceptualize the unconceptualizable system of capitalism's most recent mutation. The cyborg—sometimes animated, sometimes not, sometimes science fiction, sometimes as real as the contact lenses we wear—is an assemblage of human and nonhuman parts. To talk about cyborgs today, or even cyberpunk, however, is not unlike talking about tattoos and piercings: it is a practice that has become so conventional and tiresome that we might think we are stuck in the mid-1980s, reading William Gibson's *Neuromancer* (1984) or Donna Haraway's "A Cyborg Manifesto" (1985). But this throwback future is part of my reconceptualization of time and crisis.

So let us shift into reverse and try to remember the days when we were actually excited about cyberpunk films, those works that deal with the four C's: computers, corporations, corporality, and the city. We should now add two more C's. The fifth is that of crisis—exploding bodies, blasted land-scapes, baleful futures, and unremitting terror. Cyberpunk forced us into the heart of crisis, not into a late-nineteenth-century heart of Conradian darkness, but into a very twenty-first-century blinding light of cyberspatial implosions and system crashes. The sixth C is the new chronic, the domi-nant temporality driving the plot and characters of these narratives. I will elaborate on these recent shifts—shifts that move from cyberpunk to a new array of amnesia and memory films, but amnesia films not of *The Man-churian Candidate* sort, in which the forgetting is finally resolved by an overstuffed remembering and, as usual, an overstuffed subjectivity, but amnesia after the so-called death of the subject. This analysis then leads to speculations on amnesia after the so-called death of cinema (by an ex-tended analysis of Manu Luksch's filmless film *Faceless*) and, finally, to amnesia after the death of death itself—the contemporary moment when the chronic holds death hostage, closing the critical circle but also offering new possibilities.

Many people raced to cyborg fiction because it seemed to give a visual and a narrative face to so much cultural and social theory questioning the autonomy of the subject and the boundaries of the human body itself. But at the moment when the problem of the subject was raised in these works, when the "I" could be radically questioned, another problem was emerging that had yet to find an appropriate language: a globalized, geopolitical system in which nations were slowly losing their sovereignty. The cyber-punk form in general, therefore, was less about narrating the breakup of the subject (that is, less about speculating on new forms of identity) than it was about allegorizing the breakup of the nation, a desire for which that could still not be conceptualized or named at the time.

But there was also a crucial limit in these cyborg narratives from the 1980s and 1990s. For example, although the cyborgs necessarily exceeded the human subject in terms of the technologies of the body, their ideo-logical investments (in terms of gender identification and sexual desire, in particular) were still deeply rooted in the human. Many cyberpunk films, therefore, were effectively allegorizing our own moment in which the world-system was radically transforming in terms of technology and the political-economic "body," while its subjects' cultural and ideological in-vestments were moving at a slower speed. The narratives were figuring the unevenness between the economic effects of globalization and our experiences of this condition. They registered our inevitable reliance on older forms as we tried to produce new images, new maps, and new languages so as to make our way through a system so immense that it could not be grasped by the usual categories of perception upon which we depended.

But today, the issues of globalization and the transformation of the nation-state are more readily discussed. It is simply standard now to accept the idea that we are in a different stage of capitalism called globalization. Although this stage shares many rock-bottom rules of capitalist accumulation, it marks a transformation in the operations of the nation-state and a radically heightened state of commodity fetishism. And if even these traits do not persuade the most skeptical among us, those who see very little sociopolitical difference between today and the beginning of industrialization, then there is the unmistakable quality of the ideologies of globalization to confirm its existence, the ideologies that, with the mere naming of this condition, effectively preclude the consideration of capitalism as a problem ("globalization" is precisely that category that usurps "global capitalism" as an object of critical study) and successfully preclude the thinking of what comes after it. Either way, we have finally named the system.

And it is just when the analytical is catching up with the imaginative, when the cyborg becomes just another tired trope, just another lackluster adaptation of a Philip K. Dick novel by the executives at DreamWorks, when a new problem emerges: What do we do about this global thing that we named? How do we organize? What about the old "reform" versus "revolution" question? Whereas earlier cyberpunk films revealed something about the emerging composition of the world-system, this new strain of amnesia and memory films I will discuss teaches us something about the composition of ourselves in relation to this world-system. Or to come at this from the other direction: Now that we have named the system, is it time to rethink ourselves as political subjects? Or even to rethink the very category of political subjectivity?

To explore these questions of subjectivity, let's begin by thinking of some of the classic memory films, such as John Frankenheimer's *The Manchurian Candidate* and Alfred Hitchcock's *Spellbound*. What is at stake in these works is a certain notion of guilt and accountability. Remember the Manchurian brainwasher's words about his brilliant scheme: "We take a normally conditioned American, who's been trained to kill and has no memory of having killed. With no memory of his deed he cannot possibly feel guilt, nor will he have any reason to fear being caught, and having been relieved of those uniquely American symptoms of guilt and fear, he cannot possibly give himself away. . . . [H]is brain has not only been washed but dry-cleaned." When Frank Sinatra's character Bennett shakes Raymond out of his hypnosis and into a remembering, Raymond goes on to kill his father and mother (who were part of the conspiracy), and finally, out of guilt, himself.

In *Spellbound*, Gregory Peck's character thinks he is a famous psychoanalyst who has written the bestselling book *The Labyrinth of the Guilty Mind*. The book delivers the following watered-down Freudianism: Peck's amnesia and admission to a murder he did not commit are due to his repression of the earlier trauma of accidentally killing his kid brother. In fact, it was the former head of the psychiatric hospital who committed the murder and, by exploiting the Freudian logic, attempted to attribute the crime to the amnesiac. Whereas in *Candidate* there is the erasure of guilt to aid in a forgetting, in *Spellbound* there is the production of guilt to produce the same effect. In both films, the end of forgetting and the solving of the crime occur only when guilt and accountability are restored to the subject and no longer exploited by the criminal.

In these films, in which those committing the crimes (or those who are thought to have committed the crimes) suffer from amnesia or are hypnotized, the desire for the perfect crime, the elegantly assembled murder that cannot be pinned on the perpetrator, is expressed. Now let us think about the more recent film *Paycheck* (2003), in which a corporation employs a man to perform a service of high-tech larceny and then cleans his memory so as to preempt him from exposing any wrongdoing, to keep him from allowing his guilt or fear to get in the way. This "cleaning capacity" is a license to print money, since those who cannot be held accountable commit the crimes. But something similar is going on in terms of today's corporate logic. For example, DreamWorks, the production company of *Paycheck*, employs famous directors and stars to make big budget films that are all but guaranteed to make money—even if one of these films loses at the box office it is sure to regain its losses in international distribution and DVD sales. The circuit has been closed. The dream of *Paycheck* is the reality of DreamWorks. But this is also the reality of global capitalism.

More and more, we are grasping that the perfect crime of capitalism is built right into the system itself and scarcely needs to be hidden or disguised within false rhetoric. But then why are so few interested in any type of systemic critique? Part of the problem is that an older form of capitalist ideology is getting in the way. Even if we try, many of us cannot help but take the system's pharmaceuticals, wear its sweatshop-made clothes, and use its products and services (of course, this primarily refers to the minority of the world's population who actually have access to these goods and services). We can cognitively map the system and learn where our coffee comes from, how our shirts are made. In the end, however, we cannot avoid transgression. In an earlier Fordist moment of capitalist production, workers were bought off on the level of desire (tempted more by what type of upholstery to outfit their new cars with than by organizing revolution on the factory floor). Today, it is more about being bought off on the level of conscience, since it is impossible within commodity culture to be clean. Perhaps we no longer experience the old Fordist desire, with its accompanying moment of forgetting ("I would really like to have that commodity, no matter how it's made and how the workers are treated, and therefore I will let myself forget how the system works so that I can enjoy my new purchase"). Instead, we increasingly experience its opposite: we don't want to have that commodity, because we know (and don't want to forget) how it's made and how the workers are treated. Nevertheless, we cannot conceive of how to get by *without* purchasing it (because we see no alternative option) and we cannot prevent feelings of guilt over our participation in a loathsome system. Therefore, we let ourselves forget the vulnerability of the system precisely so that we can enjoy our purchase knowing that we could not have done otherwise. While the Fordist statement was, "I will not organize against the system because I have forgotten the system is actually rotten," today's statement is, "I will not organize against the system because I have forgotten the system is capable of being terminated." In both cases, something like amnesia or "memory washing" has occurred. But the causes and effects are qualitatively different. And again, in the new chronic moment, what we cannot make ourselves bring to consciousness is cognizance of, let alone desire for, the system's final end. However, the radical action available to us today is *not* a "waking up" to our duplicity and hypocrisy, in which we let ourselves experience all the attending guilt we are currently managing by various amnesic strategies.

Rather, the challenge is to *mobilize* our hypocrisy in a way that relies less on moral categories and more on an objective critique of the total system that has left us—via false choices—with "hypocritical" lifestyles.

There is an old conspiracy joke that goes something like this: after one hundred people are randomly called and receive the whispered message, "We're on to you, we know what you did," seventy-five of them leave town the next day. With the heightened integration of the cultural and the economic, even dreaming of a coexisting outside to the present system is a fool's paradise. The trick is not to clean our minds, to take a water-cure, to moralize against consumer capitalism, but to cultivate a principled hypocrisy—one that can justify transgression and hypocrisy based on a larger understanding of capital and the world-system.

For example, the Motion Picture Association of America is pulling out all the stops to combat media piracy and peer-to-peer file sharing. You would not go to a video store and steal a DVD, argued Jack Valenti, the association's former president, in 2003, so why do you think you can rip it off from the Internet? A few years ago, the following trailer was screened in many North American movie theaters: a working-class, African American stuntman is excited as he explains in grand detail how an action sequence was made. Then he admonishes the audience for even thinking it is acceptable to download the film for free, effectively jeopardizing his livelihood and disrespecting his labor. Likewise, the cable television industry produced a commercial involving a boy who is caught shoplifting. When rebuked by his father, he fires back the silver bullet: "But Dad, you steal satellite signals all the time!"

At this moment of new possibilities in terms of digitization, in which a rethinking of production, distribution, consumption, exhibition, ownership, authorship, and spectatorship are all up for grabs, the movie industry wants it both ways: to exploit these new possibilities and business models while ensuring it does not get burned by so many eager whiz kids deploying these very possibilities. Indeed, we can read this problem right into the films themselves, both formally and in terms of viewership.

Many contemporary amnesia films, such as *The Bourne Identity*, *The Majestic*, *Gothika*, the remake of *The Manchurian Candidate*, *50 First Dates*, and *Eternal Sunshine of the Spotless Mind* use amnesia (the radical form of the disease represented in most films is, in fact, an extremely rare medical condition) to open up possibility. Amnesia allows something to occur that the filmmaker would otherwise not be able to develop, while the disease limits this development in the name of remembering, a cure, or revenge. In this way, the fatal flaw of the perfect crime turns on the protagonist's remembering through an act of will, through a just-on-the-tip-of-the-mind instinct that is shaken into action by an unforgettable sense of morals, ethics, or love.

Such forgetting is also happening on the level of viewership. Amnesia has been a prized metaphor within film theory, most notably in terms of apparatus theory from the 1970s in which the self-erasing apparatus of the camera constructs a viewing subject. In other words, while watching a film, we forget, become amnesic, about the materiality of the film's production and technologies and thus reinforce an ideologically disempowering notion of ourselves and the world. When the lights come up, we "remember" our desire to repeat this process by purchasing another ticket and forgetting once again. The disconnection between the economic effects of our participation and the ideological effects of our viewership reproduces the film industry and ultimately dominant power relations themselves.

Today this model is dated. The forgetting of the viewer is built right into many themes and plots (*The Matrix, The Truman Show, Mulholland Dr.*) and into the very editing of the film (*Memento*). And because DVD versions of these films are always already disassembled, the relationship between remembering and forgetting is utterly confused. Still, what is remembered is an older form of capitalist ideology—that we are not at liberty to suspend the logic of profit and transgress the laws of capitalist exchange. The viewers and users internalize the guilt trip perpetrated by the leaders of industry, and by many of the artists themselves, producing a debilitating notion of hypocrisy that arrests the imagination.

There are a number of political movements that are attempting to think their way out of this logic. Among them are the antiretroviral generic drug movement, Vandana Shiva's seed movement in India, the peasant movement in Brazil, groups struggling for the cancellation of third-world debt, and groups involved with the issue of climate debt, in which the demand is on industrialized nations to pay a debt to industrializing nations for past and present actions that have contributed to climate change. These movements justify their transgressions by citing the contradictions and absurdities of capital itself, short-term health and poverty needs, by cultivating an acute historical consciousness of how the problem came into being, and a vision of alternative modes of production and consumption.[76] Of course, media piracy is not a life-and-death issue, but the possibilities for rethinking political action are great and the stakes high.

Another place to rethink these issues is in a certain strain of amnesia films coming out of Japan. In Kurosawa Kiyoshi's film *Cure*, a rash of murders has occurred in which the unlikely murderers have killed in the same way and have no memory of the recent past. We learn that there is a hypnotic spell circulated by a young psychology student who cannot remember his past or his name. He spreads the killing suggestion by way of a form of questioning. When someone asks him who he is, he responds by asking the question back to his interlocutor. When they answer, he insists on asking again. "Who are you?" is his maddening mantra. Anyone who takes the question seriously is susceptible to a radical identity transformation. The detective who is assigned to the case must risk his own self if he wants to solve the crime. By the end of the film, the detective kills the young amnesiac, but this act changes nothing, for the worm of forgetting is spreading and cannot be stopped.

This film, and many like them in Japan, cannot but be read in terms of the national postwar project of forgetting the Japanese colonial project. And, more important, these films can be read in terms of the Aum Shinrikyo event that occurred in 1995 (Aum Shinrikyo was the quasi-Buddhist cult that released poison gas in the Tokyo subway system). Many were stunned to learn that several of the cult members were just your average next-door neighbors—well-intentioned, bright people who seemed to forget themselves and their pasts. No doubt these films are trying to come to terms with these events and the dystopia of Aum and *Cure* is random, inexplicable murder. But there is also a utopian dimension, a suggestion that we can radically transform ourselves and build collectives that are organized neither on the nation nor on exclusive group formations. The cure of *Cure* is not a remembering, but a sustained forgetting of the fear and guilt that keeps us in line, as well as a refusal to return to moralizing arguments about right and wrong that effectively lock us into a capitalist box. If morality is the site of a desire for a clean conscience, then these times seem to require a nonmoralizing account of global capitalism and a rejection of the desire to "be good." This conclusion is not a call for amorality or a lack of ethics, but for an analysis of global capitalism that centers on a more objective questioning of its logic and effects—one in which we begin by questioning our very incredulity over so much inequality and corruption.[77] (This nonmoralizing critique of capitalism is developed into a more coherent program near the end of this book's final part.)

Instead of the globalization debate driven by those who speak so piously and in so much advertising copy about the new revolution upon us, or by those who moralize against the big bad corporations and become hysterical when cell phones go off in movie theaters, or by these groups' mirror image, those who refuse to recognize any shift in the world-system over the past two hundred years, we need to move in another direction. What comes after globalization is a nonmoralizing judgment on the situation and the systemic logic of global capitalism. What comes after the cyborg are new collective bodies able to deploy such an analysis and forget any allegiance to the nation, any guilt-induced reflexes, and any addictions to a chronic temporality that returns its individual subjects to their former selves.

These amnesia films disclose a set of dreams and nightmares that could only imagine the present in terms of either a radically different future or a radically identical past. Today, however, some of the most compelling contemporary films do not rely on these emerging futures or residual pasts to imagine the present, but seem more interested in focusing—in the most banal and immanent ways—on the present condition itself and, more specifically, on the present *body* itself. At a moment when so many people use the Internet to stare at naked bodies (not necessarily for the fantasy of the pornographic professional, but for the banal realism of the amateur), some of the most interesting Japanese filmmakers are asking us to stare at the body (and the world) in similarly literal ways.

Made in the time of Emperor Hirohito's protracted death due to stomach cancer and what came to be known as the tainted blood scandal (contaminated blood leading to HIV in a number of Japanese hemophiliacs), Tsukamoto Shinya's classic cyberpunk film *Tetsuo* (1989) figured the unspeakable, that the emperor's body (both the physiological one and the political one) was not only diseased, but always already unstable and on the brink of radical transformation. At the time of *Tetsuo*'s release, there was an orchestrated silence around Hirohito's illness—no one dared utter the words "cancer" and "emperor" in the same sentence. In Tsukamoto's film *Vital* (2004), however, there is not a cyborg in sight, but an amnesic, Hiroshi, who goes to medical school and ends up dissecting the body of his lover, who recently died in the same car crash that stole Hiroshi's memory. The long takes of the dissection scenes are in stark contrast to the frenetic camera movement and quick editing that mark Tsukamoto's earlier films. *Vital*'s protagonist studies medical drawings by Leonardo da Vinci. He probes the meaning of his past and the world at the same time he probes the dead body of the woman he was to marry.

This new "clinicalness" is also the hallmark of the work of two of Tsukamoto's contemporaries, Kurosawa Kiyoshi and Miike Takashi. Kurosawa, as we saw in *Cure*, is famous for bending the horror and detective genres into a cold, detached meditation on human relations and alienation. Without a single close-up and using a signature slow-moving camera reminiscent of Tarkovsky, Kurosawa upends the usual plot lines of gruesome murders, mutilated bodies, and impending danger, leading both his film's characters and viewers into calm intensity, rather than agitated panic. Less a formalist play with genre than an opening up of new ways of responding to the visible, Kurosawa's films reveal different (and what might be politically radical) ways of responding to crisis. In *Cure*, for example, no one is too terrified, and no one responds to the fear by yelling, screaming, or retreating into the nearest closet. Rather, there is a flat acceptance, a leading-with-the-chin caution, followed by measured interest and matter-of-fact plans. And this goes for the criminals as well as the detectives, for the victims as well as the victimizers, for the director as well as the viewer.

Miike, the reigning bad boy of Japanese cinema, is probably most famous for stuffing his films full of more violence and death than any director working today. If you can think it, Miike has filmed it. But the enormous popularity of Miike's films the world over is not necessarily due to the prurient desires of his young audience. If that success were attributable only to such schoolboy voyeurism, then any number of contemporary directors would easily fill the bill. Rather, Miike shoots all the gore, all the decapitations, all the sadistic shredding and stretching and spraying of the body, with a desire not to shock or disgust, I think, but simply to see all of this on film.

This approach might account for Miike's phenomenal productivity, his making nearly sixty-five films over the past fifteen years.[78] Miike makes each film without the micromanagement of a Ozu Yasujiro or the emperor-like control of a Kurosawa Akira (not to mention without the artistic seriousness of many of Miike's peers). Rather, each film seems to be made on the fly, with the next one in mind. And this is also the model for his viewers and fans; they watch a Miike film and before they can canonize and fetishize it, they have already moved to the next one. This disposable culture does not provoke what usually goes by the name "critical thinking" and fits ever too neatly into a consumerist ethos. The point is, however, that the films do not immediately point to anything outside themselves. In other words, representation is not as central as the thing itself. The existence of the film and one's engagement with it (as producer and consumer) are more important than what the film means.

Most significant in the current work of these three directors is how their films, and in particular the portrayal of the body in their films, do not primarily function as a metaphor, an allegory, or something through which one understands the world. Rather, the film and the body are the world; they are the past, present, and future. The body of the films (and the filming of the body) provokes an experience in which one looks and responds first and foremost to the film itself, rather than to something outside the film, something that transcends the film. One cannot help but think, in contrast, about the student movement and the New Left in Japan when watching Oshima Nagisa's films from the 1960s. Likewise, one could not help but think about the corporatization of Japan when watching Imamura Shohei's films from the 1960s and 1970s. But the political import of these younger filmmakers is that the films are not that precious—before you know it, the next one has come along.

This deemphasis of representational meaning was echoed when Terry Eagleton complained about so much trendy body theory in recent times (work that overemphasizes body metaphors at the expense of real, laboring bodies).[79] Eagleton was right, but only half right. Today, the body not only produces capital through its labor, but it is capital itself through its very commodification (the commodification of its diseased state through pharmaceuticals or the commodification of its genetic information or its parts through transplantation). Therefore, bodies today become not representations of capital as a system—they are already the system, and the system can be observed when we simply observe the body, as Tsukomoto, Kurosawa, and Miike ask us to do. As it does for wheat, rice, or any other capitalist commodity, the logic of the market requires a withholding—or even a destruction—of any surplus goods (in this case, life-saving medications, which can be as basic as antidiarrheal drugs) so as not to push prices so low that they jeopardize the integrity of the system itself. Bodies, too, are not excepted from this logic. How else can we explain the economic eugenics occurring in southern Africa and other parts of the global south?

People the world over are forming a similar understanding of global capitalism—one in which all ideals are at the mercy of the larger economic logic. Such recognition has existed since capitalism's inception. It is only after the Cold War and after the well-nigh total dominance of neoliberal economics, however, that such a global understanding can flourish. My temptation is to understand this more objective assessment of how things work—this *real economic*—in relation to cultural and ideological shifts that themselves are more objective.

These ideologies will necessarily express themselves differently in different locations. Culturally, I have focused on new trends in Japanese cinema. It is not a coincidence that the most interesting Japanese films today move away from a science-fiction body, with its fantastic posthumanness and its merging with machines, a depiction in which the liberation is other, to focus on *the thing itself*, on the everyday, banal body. This move is centrally engaged with the main political questions that many are facing today: How does one imagine a new form of power, a new constitution, a new state structure out of the present structure? How do we exist in a long-term crisis without leaping to fix it in some inadequate way or try to find ways to hide from it? These filmmakers present a way into these questions, a way of posing them cinematically before they can be adequately articulated in everyday political life. New political movements are facing the same thing: they feel desperately what they want to do, but are unsure about how to go about doing it. The sense is that the crises of the day have settled in, that their management has become so thoroughly taken for granted it is impossible to build a compelling critique arguing that the whole system must be overhauled, though everyone has accepted the system is permanently sick. And activists, unable to imagine the right kinds of questions for this chronic moment, may get stuck in some inescapable questions, such as "Do we need a program, or a party?" The result can be the loss of an awareness of the world itself, the body itself. And what gets found by many is a moralizing attitude that ends up mirroring and reinforcing the cartoonish representations of the world espoused by their enemies. This recognition brings us back to these films, in which there is a calm fixation on the body, without the need for long-distance allegories, redemption, or the smuggling in of a heavy-handed position about what is being portrayed. Such narrative conventions (the "revelation" of the hidden crisis or conspiracy, the moment of extrinsic rescue or redemption, the moral lesson, etc.) belong to a previous moment of capitalism.

Nevertheless, we are desperate today for images to expose the hidden crisis, so that we can manage it, if not defer it into the indefinite future. We are desperate today for the image to save us, to save us from the horrors of the past, the crimes of the present, and the threats of the future. But it is precisely this desperate desire to be saved by the image that disables the image and turns it into a set of handcuffs, into the perfect device of repression—which is to say that our desire for the image to save us from crisis (from incoming bombs, from cancerous tumors, from the emptiness and boredom of everyday life) functions to compromise our desire and to limit the future. However much we hope for it, the image (from the image of surveillance to that of medical imaging) cannot represent crisis, because the very condition of crisis, which is the logic of the social and the body, is unrepresentable. To rely on the image to catch a transgression is to forgo an analysis of how the ultimate transgression (the very condition of transgression, that is the structural logic itself) is absent and unrepresentable and, therefore, cannot be exposed by the image.

But we must be careful when arriving at this unrepresentability of the image. As Jacques Rancière concludes in his *The Future of the Image*, the "logic of the unrepresentable can only be sustained by a hyperbole that ends up destroying it."[80] Here Rancière is arguing that the moment we give ourselves over to the sublime, mysterious alterity of a force that we recognize as unrepresentable, is the very moment that we just represented this unrepresentable force—in other words, a certain regime of representability is instantiated when we stress unrepresentability. Now we are in a double-bind: recognizing the fantasy structure of *both* the desire for representation and the desire for unrepresentability. When we evacuate the image of any representational claim on the world we are faced with the same problem as when we evacuate the world of any representational claim on the image, that is, we effectively evacuate the world and the image of making any analytic sense. The consequences of this evacuation are best demonstrated by the following facile claims, "the image exposes the truth" or "I don't trust that the image has any truth." These claims work in ideological coordination with the following ones, "the world system is totally determined by an underlying logic" or "the world system has no underlying logic." These overinvestments and underinvestments in the truth of the image and the logic of the world meet at the same point of critical surrender and today form the reactionary core of most aesthetic and political positions.

A FUTURE WITHOUT AN IMAGE

How, then, can we liberate the image from our desires and our desires from the image? And how can we pursue this freedom when we are so driven to be saved by the image?

In 1995, in coordination with the one-hundred-year anniversary of cinema, Yoshida Kiju was asked to make a film with the original cinematograph camera used by the Lumière brothers at the end of the nineteenth century. Yoshida, one of the key filmmakers of the Japanese new wave in the 1960s and 1970s, chose to film Hiroshima by not filming it. He set up the camera right in front of the famous Genbaku Dome, one of the only structures left standing in the area where the first atomic bomb exploded on August 6, 1945, and made a short film about the impossibility of representing the event.

The film begins on a bullet train pulling into Hiroshima, followed by Yoshida, the director, fiddling with the cinematograph. He cannot fit the film properly into the wooden box. "Impossible, impossible," Yoshida exclaims in French to an assistant, a French man who presumably knows how to run this relic of a camera. He then faces a camera and explains (now in Japanese) that today people think that the cinema, and images in general, can explain everything, everything in the world. But this is not true. According to Yoshida, cinema cannot show anything: "And this is what I'll try to show you." Here we get our first paradox: How can you show something that cannot be shown? Yoshida will show what is at stake in our desire for cinema to show us the world.

Before Yoshida finishes speaking, the scene shifts to a piece of film stock moving right to left across the frame. The film stock for the original cinematograph was composed of sixteen frames per second. Each frame is separated from the next by a small, empty space, just as they are in standard 35-millimeter film that runs at twenty-four frames per second. And this space in between the frames is unrepresentable by the projector when the film is moving at the proper speed. But this negative space (or what might be called *ma*, the interstice, in Japanese) is the condition of cinema,

An image of film stock moving across the frame in a project directed by Yoshida Kiju.

without which there would be no images at all. The structure of presence and absence, of possibility and impossibility, is built right into the most fundamental unit of cinema—the film frame itself.

This argument is echoed in the following scene, when we see Yoshida and the French assistant standing behind the cinematograph but also being filmed by it. This takes us to the second paradox: How does one simultaneously film with the camera and film the camera itself? How does one use and draw attention to the camera or, by extension, any apparatus, even our language and concepts? How can we think about or desire something (Hiroshima, cinema) and engage the very ways by which we think about or desire it? There is a blind spot that cannot be overcome, and Yoshida argues that we must examine our desire to expose or hide this blind spot—a desire that is buried in the very act of watching.

This is followed by a slow wipe revealing what Yoshida and the cameraman are looking at: ground zero of the Hiroshima bomb blast, the dome. But there is someone casually walking by it, even though we hear the sound of the bomb blast. There are several times here: the time of the camera (1895); the time of the bomb (August 6, 1945); the time of Yoshida filming this (1995); and the time of our watching (our present). One reason

Yoshida and his assistant stand behind the cinematograph while also being filmed by it.

cinema cannot represent the world is its inability to show us this multitude of time; the film stock runs through the machine at a regulated speed. The third paradox, then, is how to represent the simultaneity of these different paradigms of time, how to catch the way we live in various times and how these various times live in us (at the same time), how to catch this echo chamber of time with a technology that moves at a single speed.

This is followed by a cut to a close-up of the faces of Japanese men and women (in the present). They were children of the bomb and are now about sixty. Yoshida asks over the shot of these faces, can cinema describe what happened at that instant? And here we get the force of the idea, and our next paradox: if Yoshida and the camera had been filming when the bombs were dropped, he (and the camera) would have been obliterated. Filming the spot of the blast from ground zero necessitates one's own annihilation. Whereas the second paradox emphasizes time, the third emphasizes space, how the actual location that one is filming (and from where one is filming) shapes what can be filmed.

Yoshida wants to engage these paradoxes, these impossibilities, by way of film. He argues that we should not prescribe what the image can do, nor should we presume that it can resolve these paradoxes (by technological

Yoshida and the cameraman watch a serene postwar scene of the Genbaku Dome as a bomb blast is heard on the soundtrack.

advances in imaging, including everything from satellites to hidden cameras). This presumption, this desire invested in the image, is dangerous and as potentially explosive as the dropping of the atomic bombs. But there were images taken of the blast from the *Enola Gay*, the U.S. plane that dropped the bomb on Hiroshima. This hard fact takes us to surveillance and to the desire to be both connected and disconnected to what is being filmed. Military images. Police images. Private security images. Satellites. Drones. Hidden cameras. The dream is that every inch of every space on the planet is filmed and stored. If we can produce images of every place in real time, so the desire goes, then there is nowhere left to hide, thus crime and corruption are eliminated.

In 2007 Manu Luksch released the film *Faceless*, a film composed entirely of images recorded by closed caption television (CCTV) surveillance cameras in the United Kingdom. Britain is famous for having 20 percent of the world's twenty-one million CCTV cameras, making the country the most surveyed, with the average person caught on film three hundred times a day.[81] Over a period of four years, Luksch staged events in front of designated CCTV cameras, applied under the U.K. Data Protection Act to retrieve these images, and then edited the images together into a science-

Film still from *Faceless* (2007, Sixpack Film), directed by Manu Luksch.

fiction narrative about a utopian-dystopian society in which contented citizens live under constant surveillance. Luksch is not the first to employ "legal readymades" or to stage events in front of surveillance cameras, but *Faceless* does confront the new chronic like very few art works today— perhaps because, like the concept of the new chronic itself, Luksch's film reveals the contemporary inextricability of the cultural, political, and medical.[82]

The fifty-minute film begins with the following text: "A new time / the new machine / amidst which survives / a singular dream." "New time" refers to a mode of temporality that organizes the future society of the film, one that has no past and no future, just a perpetual present. And the new machine broadcasts "pulses of real time" and monitors the data traces left behind by every citizen. The new machine also dispatches overseers to correct "any errors and deviations instantly." A voiceover explains the reasoning for all this, "In a distant era, people had become discontented. Anxiety about the future and guilt over the past caused great unhappiness. The present was continuously in short supply. Then a reform of the calendar was proposed to dispense with the troublesome past and future and fill everyone's lives with the perfect present. An advanced technology called

Faceless was composed entirely of images recorded by CCTV surveillance cameras in the United Kingdom.

the new machine was developed to supersede the past and future. And soon afterward the system of real time was unanimously accepted."

With no past to provoke guilt and regret and no future to produce fear and anxiety, the citizens of real time are happy: "The perfect and perpetual present is the heartbeat of the healthy universe." These subjects, who live in this perpetual present, are faceless, since it is the face that betrays the various emotions tied to the past and future. The main character, played by Luksch herself, however, is haunted by a memory, despite the social fact that memories no longer exist. This dislocation causes her nausea and also the traumatic moment in the film when she sees her own face. Assisted by a group of spectral children (themselves outside the discipline of the new machine) and a man, the protagonist's former husband, who sends her a letter, the woman attempts to break out of real time without being caught by the system's overseers. She does, but only to then doubt whether her newly discovered dreams and past are in fact her own. Perhaps, she supposes, they are nostalgia for a time that never was or a prison of another perfect present. The ambiguity of the end is the answer to the woman's questions, for doubt is the single quality that cannot exist in the dystopia.

At the end of *Faceless*, there is a title explaining that the images were attained under the terms of the Data Protection Act and that in compliance with this legislation the privacy of all third-party persons was protected. The data controllers are required to obscure the faces of all those in the data footage (usually with black ovals), only leaving visible the face of the data subject. These protocols are what shape *Faceless*'s narrative, a narrative about a woman who has a face amid the faceless. The last line of the film explains that "the plot evolved alongside the process of obtaining the recordings," and Luksch and Mukul Patel, in a separate text, explain that many of the requests for footage were disregarded due to a combination of avoidance, incompetence, and technology failure. The story, therefore, had to be continually rewritten.[83]

With surveillance, the desire (of the state, our desire) is to catch the criminal red-handed. There is a desire to reveal a transgression, to get at the truth. The state wants to catch our crimes, and we want to catch the state's crimes. When we complain, we are reassured that if we are not committing any crime then we have nothing to worry about. And when we catch those in power transgressing, the state apologizes and promises to clean up. But the problem is that crimes are committed even when there is no crime to catch. Crime is built right into the system itself. The desire to catch the truth of the situation will always come up short, because the truth of the situation cannot be caught.

We have already established this argument when referring to crisis and capitalism. The inequality built into capitalism is not simply due to the pursuit of so many bad capitalists, but to the structure of relations that are reproduced daily with the making and selling of commodities—especially when the capitalist is benevolent. To criticize capitalism based on its transgressive acts, therefore, is to forgo an analysis of how capitalism produces inequality when it works exactly as it is designed to work. In the context of surveillance, to rely on the image to catch our transgressive acts is to forgo an analysis of how the ultimate transgressive act cannot be exposed by the image. It is true that the surveillance image can show what is absent (that the accused was not present at the scene of the crime), but the surveillance image cannot show an absent presence. Nor can it show what is presently absent, but what can come into being in the future. Just as the unconscious is not something that can be exposed like an X-ray, so too is the logic of the present and the possibilities of the future unimageable.

Contemporary medical imaging attempts to overcome this limit by not only imaging the future development of a tumor, but by imaging genetic material that can indicate whether a tumor that does not exist in the present might develop in the future. Philip K. Dick's short story "Minority Report" (1956), about a future society in which criminals are caught before they commit their crimes, has something to say about this desire to predict the future. In the story, a detective receives information from three "precogs" who are able to read images of crimes before they happen. This information is then used to preempt the crime and sustain the perfect society. The easy critique of such a society is to stress the fallibility of all totalizing systems of surveillance. Indeed, this is the direction taken by the narrative of the film *Minority Report* (Steven Spielberg's film adaptation romanticizes this human fallibility, while Dick's story remains true to the dystopia). But we might want to go further and consider what is problematic about such a society when the surveillance technologies work to perfection? In other words, what are the consequences when the system works rather than when it fails? What is closed off when the system reduces the future to the present? New fantasies of surveillance and medical imaging cultivate a desire to manage or even preempt crime and illness in a way that foreclose alternative ways of organizing society and dealing with death. By placing so much desire in the image to save us, we repress our unconscious desire for a radically different future—a future without an image.

Faceless, like *Minority Report*, encounters one of the classic paradoxes of dystopian (or anti-utopian) fiction, which is how to criticize the false utopia of the state (usually one that attempts to extinguish any semblance of the human) while not effectively sanctioning an earlier "more authentic" moment and therefore legitimating all of that moment's violence and injustice in the name of human feeling. The countless anticommunist fears that circulate in capitalist societies, fears about total control, grey suits, and one-dimensionality, reveal how these fears legitimate so much crime and inequality and one-dimensionality in return. In *Faceless* this paradox takes the form of the painful human emotions that are enabled by the former calendar, a past when surveillance was not prevalent, by a relation to the past and future. Is it possible to be free from guilt, regret, fear, and anxiety (or at least weaken these emotions' debilitating effects) without returning to the previous, harmful temporality? And if not, then might the type of unfreedoms of "real time" not be a sensible trade-off? To put this paradox another way: the quality that drives the protagonist of *Faceless* to question the new machine and real time (her sensitivity and excessive feeling) is the same quality that will stop her from crossing over to the new.

Chris Marker confronts this paradox when the protagonist of *La jetée* (1962) chooses to return to the past and to the power of his memory of that time. It is only through the power of that memory that he is chosen by his captors to partake in time-travel experiments and is able to receive from the future a power pack that will reboot human civilization. But it is also the power of memory that forces him to reject the invitation from those in the future to join them. If he were to join the future instead of returning to the past, he would not be able to get to the past in the first place—his drive has to be so strong that he would then not be able to give it up. It is for this reason that these utopian-dystopian narratives are always at the same time narratives of sacrifice—since the protagonist can only open the space for someone else to enter, someone whose drives are organized differently.

At stake in these narratives is the way the contrast between the human and the nonhuman are invariably represented. The dominant representation has it that humans in free societies feel, while nonhumans (and humans in unfree societies) do not. It is at this point that we should historicize the various human emotions and argue that it is human to feel regret or guilt or anxiety, but there are qualitatively different modes of regret or guilt or anxiety. There is capitalist guilt and capitalist anxiety, which are not the same as feudal guilt or socialist anxiety. Perhaps this explains the popularity of Soviet-bloc Cold War jokes in the West right after the end of the Cold War. The freshness of the jokes to a Western sensibility (jokes involving this or that comrade followed by the KGB) is not due to the fact that they are being heard for the first time, but due to the different quality of the emotions. The most interesting dystopian fiction effectively reveals and criticizes the key reactionary ideological assumption of our moment: that the emotional qualities that make us human remain consistent throughout history, that if we now feel we could not be happy in some imaginable, different society, then the construction of that society should not be pursued. Even if we are correct—even if we could never attain what we now call happiness in, say, a communist society—this is neither an argument against communism nor an admission that communism has to be grim. We will not have our current version of happiness, but we will have some other kind of happiness. We may not be more or less happy, but our happiness will be of a different order altogether. By

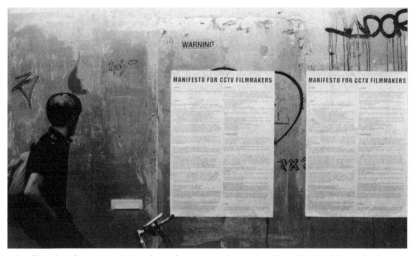

The film also features a "Manifesto for CCTV Filmmakers" established by Luksch.

critiquing this false naturalization and eternalization of our own emotions, therefore, great dystopian fiction is always a thoughtful reflection on time, or how to account for change from one historical moment to the next.

In *La jetée* we have a single temporality occurring at different times, while in *Faceless* we have different temporalities occurring at the same time. If it were not for the formal experiments employed by both Marker and Luksch, their films would end up more trap than wormhole. Marker uses the form of *La jetée*, the photographs, as a way to open up the gaps in this all-enveloping temporality—to break into the circle of time. Likewise, Luksch makes something close to carbon-free cinema. She does not use her own camera or her own video. But the work still desires to be a film, clearly demonstrated in one of *Faceless*'s key titles: *A Film by Manu Luksch*. Luksch wants to kill film while retaining the category of film, or at least the category of expanded film.[84] Indeed, this links up with Yoshida's desire to show us the limits of film by way of film itself.

This is also evident in the way *Faceless* includes a manifesto, a set of principles and instructions in which Luksch establishes how to make film that intervenes with the surveillance state. Unlike the real time of *Faceless*'s narrative, the manifesto is a form that invokes the future—it is about anticipation. It is a "futuristic speech act," as it is called by Martin Puchner in his *Poetry of the Revolution*. Puchner explains that in the modernist moment "the manifesto, now situated both in politics and in art, became a genre through which art and politics could communicate, a kind of membrane that allowed for exchange between them even as it also kept them apart."[85]

The way the manifesto links and delinks theory and practice is crucial to its political potential, since this theory of practice—and practice of theory—mirrors the way the present and future relate to each other, how future potential is already integrated into the present situation. Puchner writes that there is "a remainder of theatricality in the manifesto that allows it to speak in the theatrical mode 'as if,' as if the unified proletariat already existed."[86] Just as something between theory and practice must remain uneven for the radical aspect of the manifesto to work, something between the present and the future must also remain uneven for a radical politics to thrive.

What differentiates our current moment of the manifesto from the past one that Puchner analyzes, however, is that the "as if" has shifted to the "already." The unified proletariat already exists, meaning that collective action exists as the only revolutionary possibility (and here we would need a properly capacious definition of what a late-capitalist, global proletariat is). We do not need to act "as if" the unified proletariat exists (to talk ourselves into the courage required to pursue revolution), but to recognize that it already exists as the only effective challenge to the new chronic. Today's manifesto is more a philosophical-political statement about the truth of an already existing future than a theatrical one about the wish for something that does not yet exist. Just because something already exists, however, does not mean that its effects are realized. Indeed, just because we are already dead does not mean that we have died. Unevenness still exists between the present and the future, but now the present has overtaken the future. How to claim the radicalness of this reconfigured future, of this future that has already happened, is the most crucial political, cultural, and theoretical question of our current moment.

PART 2

The Global Abyss

Not long after 9/11, my family doctor called with his own bit of terror. My white blood cells were through the roof, which meant only one thing: leukemia. "Your body's in crisis; get to the hospital at once." It was early evening and I was at home. I focused on the color of the walls and on the shape of my computer screen, resolving not to rush, not to leave at once. If the rest of my life was going to be about the spectacle of cancer, about all the high-stakes decisions that would need to be made regarding therapies and wills and serious phone calls to friends and family, then I was going to take my time, linger in my last few moments of the banal, in the everydayness to which I could never return once walking out the door, hailing a cab, and admitting myself into the Canadian health care system.

Two years earlier, I had left the United States to come to Canada as a professor at the University of Toronto. I had a good job at a university in the United States, but decided Canada would be a more stimulating place to live and work.

Toronto was livable, a remarkable film center, and one of the most impressive diasporic destinations in the world. I could not think of a better situation—plus, leaving the United States was appealing. I was tired of the inequality, of the lack of any national humility, of the browbeating foreign policy and prevailing ideology that yoked together personal wealth creation with freedom and democracy. Even the mainstream left was giving up on the welfare state and whatever remaining space there was to think critically, to question assumptions and play with ideas and alternatives, was quickly being snuffed out. I did not believe Canada would be significantly different, but at least it would be saner, a place where I could walk down the street and look into the eyes of others without class and racial tension informing every glance. How bad could it be when even the Canadian right believed in socialized medicine?

I was in perfect health. Like most incoming faculty members, I was granted a one-year work visa. And because I was a U.S. citizen, the North American Free Trade Agreement (NAFTA) made it even easier to work in Canada. The next year, I extended the visa while preparing my application for permanent residency. But then, only weeks before the university lawyers were to submit my application to Immigration Canada, came the leukemia diagnosis. I was feeling fine, without a single symptom. The lawyers told me to hold off on the permanent residency application, for now I had a preexisting illness—a fact that would change everything. I remember thinking that the lawyers were mistaken, that I did not have a preexisting illness, that I came to Canada two years earlier in good health and at the request of the university. All this could be proven. But at the time my immigration status was the least of my worries, so I followed the lawyers and their reasoning, which was clear if not brutally elegant: I would either die and there would be no reason to care about my immigration status or I would be cured and would apply when all of this was behind me. Even I, the unfortunate subject of such unforgiving logic, could laugh at its legalistic beauty. There was no reason, therefore, to submit an application before finding out my fate—besides, I might be found medically inadmissible if I applied while undergoing treatment.

As usual, however, things are never so simple, never so beautiful. Today I am in remission, due to what has quickly become one of the most successful new drugs in the history of medicine, Gleevec, a genetically modified oral medication that targets the chromosomal mutation, binds itself to it, and manages the disease without severe side effects. The mainstream media and numerous support groups might still characterize cancer as the big "C," a cartoonish enemy to be conquered, beaten, fought with high hopes and an unwavering spirit, but for medical researchers, clinicians, and pharmaceutical executives cancer is transforming from something to be cured to something to manage, like arthritis or high blood pressure.

Gleevec is expensive but fortunately paid for by my private, university health insurance. After going into remission, I applied for permanent residency. I was still concerned about being found medically inadmissible, but when researching what this meant it seemed clear that I should pass through without a problem. Medical inadmissibility in Canada means that an applicant for permanent residency will place undue stress on the health care system, which translates into costing the system more than an average Canadian would over a period of five years. Since my drugs were privately funded and I only saw a specialist about three times a year for a basic checkup, my cost to the Ontario Health Insurance Plan (OHIP) was well within the prospective amount expected by any other resident or citizen.

Following 9/11, and the fresh realization that immigration in the United States and Canada could not be separated from security concerns, the wait to hear about permanent residency applications grew significantly longer. In the meantime, I recommitted to Canada, never second-guessing my decision—until, that is, my application for permanent residency was rejected. After misspelling my name and completely misrepresenting my situation (referring to the wrong form of leukemia and treatment), the designated Immigration Officer ended her short letter with the following phrase: "Thank you for the interest you have shown in Canada." I was now expected to go home (without the opportunity to visit Canada again) as if the last six years I had spent in the country had only represented a "passing interest." I was to give up my job and my life in Canada, and return to the States. I was being treated like a felon, one whose crime was getting cancer at the wrong time.

What started out as so much bad luck, like winning the lottery in reverse (the type of accident you shake your head at, maybe even laugh off), quickly turned into a real worst-case scenario. I had been kicked out of Canada. Since I did not have a job waiting for me in the United States at the time, and since health care in the United States is privatized, I was being forced to return with no employment and thus no access to my medication.

Calling this a death sentence, I think, is only a slight exaggeration. It is a rather singular situation. If I had been Canadian and the United States had rejected my green card application, I would have at least returned to proper health care. Perhaps my situation was more on par with that of someone with HIV who must return to his home country, where adequate anti-retroviral medications are not accessible. The letter from the immigration officer was fraught with errors, insisting that I was receiving chemotherapy that was expensive and publicly funded. How could there be such an egregious error (not to mention such careless spelling and grammatical mistakes) in such a deadly serious piece of government correspondence?

When I would express such incredulity to my lawyers, they would say that this is to be expected. Somehow the fact that I was not alone in being treated unfairly and that I had entered some bureaucratic no-space was supposed to make me feel better. It did not. But it did disabuse me of any residual liberalism (however unconscious), any hope that the system is not structured in dominance and that somehow, somewhere reason would prevail, the belief that a human being would step up and recognize my situation and do the right thing. Unfortunately, a human being did step up, but she wrote and signed a refusal letter that was so poorly written and patently flawed it would embarrass even the most dispassionate reader.

Despite all of this, I still felt confident that I would receive my permanent-resident status. And if I did, I was certain the bureaucrats would then argue that, in fact, the system worked, that the final reversal of my refusal was a testament to how exceptional cases finally get acted upon with human compassion and care. I reject this. The system *did* work—as evidenced by how it fumbled my case. And even though I did finally receive residency, I stand by this critique, since my case should not have been admissible within the system's own logic. On the day before my appeal was to go to court, the minister of justice called to ask if I would accept a settlement in which my application would be judged without any reference to my medical history. The timing was connected to a key supreme court case that had just been decided, *Hilewitz v. Canada*, a case involving immigration and illness that might favor my case. I discuss *Hilewitz v. Canada* below in the section called "National Sickness and Global Health."

What does it mean to argue that the system worked or did not work? For that matter, what is this thing we call the system? When and where does it begin and end? How does it connect to other systems? And what role do humans play in relation to it, as both creators of its logic and victims of its effects? These questions and my experience open up a set of problems that reveal some of the most important contradictions of our present historical moment. I was free falling in this no-space, in this legal crack cum global abyss that has recently broken wide open as the world-system reconfigures and institutions, corporate, national, and ideological, struggle to keep pace.

A new global abyss has opened up precisely in this gap between ever-shifting economic and political realities and the institutions and ideologies we have available to us to cope with these new realities—institutions and ideologies suited to another moment of capitalism and unfit for the present. With transformations in immigration, labor flows, geopolitics, health care, biotechnology, media, family, views about life and death, and the future, as well as the operations and experiences of power itself, new spaces have emerged in which some of us—however fortunately or unfortunately —fall. Of course, I am not alone and do not represent the most dire of cases, but this new predicament, this global abyss, is what a growing number of us are compelled to inhabit.

These spaces come in between our present institutions, in between our categories of thinking; they are unimaginable without the momentum and tumult of the freefall. In what follows, I will map out six major shifts that constitute the global abyss and account for why our institutions and thinking break down. As I mentioned in part one, the disagreement over the status of the nation within the debate over globalization, whether it is still as strong as it once was, is a false problem. This is primarily because the various levels of any social formation move at different speeds, so that one is compelled to consider how national culture, for example, might be out in front of or lagging behind economic developments within the nation. This nonsynchronous quality is intensified, moreover, by a similar unevenness among the various levels of the global system. Right now we live at a moment when the nation-state and the global system are banging into each other, leaving in their chaotic wake not only such bizarrely unlucky cases as my own, but a rather unique space for change.

It is the smooth but false solidness of law itself that is at stake. Some very important transformations are under way in the immigration law governing who is allowed into a nation, the economic law of expansion and profit, the biological law regarding what constitutes illness and cure, if not death itself, and the law regarding what counts as family and legitimate love (including the more unspoken laws of cultural affect and the unconscious experiences of power). These transformations have yet to find an appropriate language, save the politically transparent and expedient languages trumpeted daily by our media blowhards. Is it any surprise that the world seems to resemble the cynical and facile cartoon in which good and evil spar, rather than the more objective historical reality of a world structured in inequality? Is it any surprise that we keep insisting that the world is driven to violence by our worst human instincts, rather than accepting the more obvious explanation, that such violence is driven primarily by the logic of our modern social systems?

Righteous indignation comes most readily when one is innocent and yet still accused of breaking the law, or when one knows deep in her being that she is being wronged, not by mistake, oversight, or false accusation, but by the very laws, self-evident truths, and benevolent constitutions that represent so much hope and development in the modern world. In other words, righteous indignation comes in two forms: first, when the whistleblower reveals the politician's hand in the till or the cop's finger too fast on the trigger; and second, when those in power acted to the letter of the law and we were still sold down the river. It is precisely this second form of righteous indignation (one that resists moralization while pursing a more systemic critique) that is becoming the most prevalent and the most revolutionary political consciousness the world over. How it is expressed, in either its various fundamentalist forms (so dangerous and violent) or in more radically reconstructive and just ways, depends upon how we experience the global abyss and give ourselves over to the freefall. In this section, therefore, I analyze what the global abyss opens up for intellectual thought and political action today. Only after establishing this new space and how it relates to the new chronic time examined in the previous section, will the stage be set for the introduction of "the already dead."

Labor Flows and Immigration Fears

Radical unevenness structures the global abyss. Different levels, different logics, different energies smash into each other, generating seismic waves strong enough to dislocate whole populations, or, at the very least, utterly wreck individual lives. The most unambiguous fault line of the global abyss is located on the national border where so many migrants cross. When I crossed into Canada, odds were that I would be unshaken by the new formations—just another gratuitously chosen statistic overstating the smooth transitions of the global economy. But I had fallen into the abyss. Pulling on one side was the university and business community that needed foreign labor (flexible and mobile) in order to stay competitive and innovative, while on the other end was the ministry of immigration (rigid and static) that had to minimize its outlays and retain the integrity of the nation-state, an integrity based on exclusive entitlements and legal status. My experience is not too dissimilar to those of migrant workers, the hard-working, undervalued, and vulnerable laborers essential to late capitalist economies. The U.S. economy is dependent on this alternative labor pool, this brown-bodied, don't-ask-don't-tell market filled with farmhands, nannies, construction workers, caretakers for the elderly, and those cleaning up after the professional labor pool. But the already denuded welfare state would fold if social security and proper benefits were adequately granted, especially as taxes in the United States are at such historically low levels. The one untouchable position for any politician in the United States, more than going to war, cleaning up the environment, or feeding the hungry, is raising taxes. It would be much more honest to recognize that the insourcing of undocumented workers (this informal economy) is driven more by the necessity of free flowing capital, goods, and workers than by the desperate poor coming to the land of milk and honey.

But an analysis stressing the sober and structural logic of the global economy over the Hollywood script of desperate individuals fleeing corrupt third-world countries is virtually impossible under the soggy weight of globalization-speak, a language and way of understanding the world not unlike the language of modernization in the nineteenth century and the twentieth, a language in which everything and everyone steadily progresses, grows more effective, and richer if, and only if, everyone bows to the shrine of commodity production and profit creation. With its compliance to the same uncontested rules, modernization theory holds that we all move forward together, despite the fact that the adults (Western Europe, North America, and sometimes Japan) move faster, as the children (pretty much the rest of the world) chase behind. What the last forty years have taught us, however, is that the development of one nation usually comes at the expense of another. Or, more to the point, certain privileged sectors of one nation usually coordinate with certain sectors of another. What is called uneven development, therefore, is not something that occurs when the system goes wrong (something to be overcome by a more rigorous obedience to the rules of capitalist accumulation or remedied by less corruption in third-world despots), but is actually what occurs when everything goes right, when the system works as it is designed to work.

And yet even this critique of modernization theory (which usually goes by the name of dependency theory or world systems theory) should be rethought today. Globalization processes currently undercut the sovereignty of the nation-state in terms of political economy, while the nation's sovereignty is as stable and strong as ever in terms of immigration policy and in the cultural ways we understand ourselves as national subjects, for instance, in the way we still define ourselves in relation to various outsiders, however much these outsiders (in this case, undocumented immigrants) are located inside and represent a foundation on which the inside stands and prospers. If there still is unevenness, then it is not organized as much around nations or even regions, but around global categories of race, gender, class, and health, as well as around the various institutions that organize our everyday life.

I would like to think we all know that most migrant workers are driven more by the global economy than their local misery. But, somehow, like the slaughter of aboriginal peoples around the world, we must repeat a collective forgetting of this truth or else the hypocrisy of our everyday lives would simply be too much to bear (and this psychic logic of forgetting might very well be doubly required by the undocumented workers themselves, in order to somehow manage their common vulnerability). But this is not only about systems, structures, and capitalist logic, but real people who have basic human rights, regardless of where they were born and what color passports they carry. When Bernard Kerik, George W. Bush's selection for homeland security czar in 2004, withdrew his name due to the revelation that he had employed an undocumented Filipina nanny for years, he punctuated his apology with the following rationalization: "She loved my children and my children loved her." Unfortunately, this love (which, like most everything else under capitalism, is privatized) is only directed toward those closest to us—toward those for whom we would kill or break the law—and thus rises above everything else in order to justify and reproduce an uneven system in which our needs and constraints (personal, national, and global) are in stark contradiction. Love might always be irrational, but it need not be naive.

As for my own case, the fatal flaw was my medical diagnosis, which I imagine did not necessarily threaten Canadian health care (almost everyone will stress the health care system, given that everyone will die, with the majority dying in hospital and incurring enormous costs). Rather, the real threat is a host of immigrants who apply primarily for medical care they cannot receive in their home countries. If there were no medical inadmissibility guidelines, the current integrity of the nation-state, however waning, could be thoroughly destabilized. The threat is to the economic bottom line of the nation that cannot afford such life-saving medical expenditures —it cannot, that is, as currently configured. Since I came to Canada without an illness and lived and worked for over two years before being diagnosed, Immigration Canada could not be too worried that there would be a spate of cases like mine. Rather, the worry was that the space opened up by my legal exception would be exploited as a loophole and burden an already fragile health care system.

There are cases similar to my own occurring every day on so many Canadian farms and in so many Canadian factories and brothels. For example, there is Hermelindo Gutièrrez, a Mexican citizen working at an Ontario tulip farm every spring and fall for six years. Gutièrrez was diagnosed with kidney failure and required life-saving dialysis.[1] Unlike the estimated 50,000 undocumented workers in Canada, Gutièrrez acquired a special status under the Seasonal Agricultural Workers Program to come to Canada and work in the fields. Begun in 1966, this guest-worker program was a model for managed migration in which workers would pay income taxes toward employment insurance and the Canadian Pension Plan, while in return they would collect on their pensions upon retirement and, most important, receive a provincial health card valid for one year. As I did, Gutièrrez passed a medical test upon first arriving in the country, but after that he became ill. Once the illness was public, the Mexican consulate tried, under the pretense that Gutièrrez had a preexisting illness, to repatriate him before he grew worse. The Canadian government held the same position and effectively supported the Mexican line.

Is Gutièrrez's kidney failure due to his work on the Canadian farm? Probably not. But as those who study the etiology of illness know, there are many pathways and triggers that lead to the manifestation of a disease (the genetic, the environmental, the physical, and psychological, among others, making it nearly impossible to pinpoint a pure point of origin or cause). Regardless of what led to the illness, or the moment at which it emerged, Gutièrrez will surely die if he returns to his Mexican village where there is little access to the required medical care. Despite this tough break dealt to Gutièrrez, it is quite legitimate to argue that Canada cannot be held responsible for the inadequacy of the Mexican health care system. And it might even be legitimate to argue that Canada should not be on the hook for Gutièrrez's specific case. But there is something larger going on here that exceeds the differences in health care offered by Canada and Mexico.

This has to do with the way the fragility of a national health care system is inextricably linked to the fragility of the national welfare state. For example, in what is called the "Respondent's Memorandum of Argument" I received from the ministry of justice regarding my case, the government lawyer made the following point: "The Respondent [the government's lawyer] notes that while at first blush it appears that the Applicant [me] is currently 'privately' paying for his medical services, that members of the Canadian public are ultimately paying for his medication through the rates paid to the insurance company from a publicly-funded University."[2] The most crucial dimension to this argument is the distinction between public and private. Is it really so easy to distinguish between the two spheres today? In the United States, so-called public universities (such as the University of California and the University of Michigan) receive more private funds from for-profit corporations than most private universities. While a private university, such as MIT, is one of the top recipients of public funds, especially from the Department of Defense. It is true that Canada does not have any private universities, but it is also true that the University of Toronto receives, by far, more private contributions than any other Canadian university. In fact, during its last fundraising campaign the university accumulated over $1.2 billion from private donors and corporations, an endowment target that was met one year ahead of schedule. Not surprisingly this vigorous campaign came precisely at the time when Ontario's conservative government imposed a 15 percent reduction on the higher education budget.[3]

Former University of Toronto president Robert Birgeneau accounted for this new reliance on the private sector as a protective measure against the capriciousness of government spending. "Fundraising hasn't made the

University of Toronto more vulnerable," Birgeneau argued, "if anything it has made the institution less vulnerable to the ups and downs of government funding."[4] The assumption here is that nation-states are no different than private corporations. But as the United States demonstrates on a daily basis, governments can engage in deep deficit spending as long as confidence in them (and in the larger capitalist system) does not drop too much, or as long as there is the necessary political might to back up the most reckless economic policy. How else can we explain the fact that when the United States betrays the tenets of neoliberal capitalism barely anyone raises an eyebrow, while when any number of less developed countries step out of line, the IMF clicks its tongue before imposing even tighter restrictions and austerity measures? The point is that private corporations operate under a different logic than nation-states, for they must report first and foremost to their shareholders. Although it appears as though nations can go bankrupt just as corporations can, in actual fact this is not the case. While there may be national defaults on loans and other provisions for private corporations, the nation's citizens have no choice (short of death) but to continue living, to continue working to sustain their lives. They cannot, in other words, simply skip to a more solvent nation to find work (at least they cannot do so in the formal economy, but this is precisely what fuels the informal economy). The worker of a bankrupt company, in contrast, must endure unemployment until finding work in another company.

What Birgeneau failed to recognize is that the reliance on corporate funding is qualitatively different than the reliance on government funding, for the rules of the market are contingent on different principles. This slipperiness when employing the private and public is at the heart of contemporary globalization processes. To simply tiptoe around this slipperiness, however, is not enough, for the ambiguity over these two realms was already built into the binary when first configured during the emergence of market capitalism back in the sixteenth century. For example, the modern public sphere was originally conceptualized as a space to generate public opinion and democratic participation—always with the explicit purpose of reproducing the power of the nation-state. Idealized as a non-hierarchical space of open access, what occurred in a salon, café, or reading group must serve as an essential check on state power as well as a key instrument in shaping the policies of the state. Criticism of the public sphere, therefore, invariably turns on who is excluded from such a key collective space, the inequality among those participating, and how the space is in fact compromised by both state-driven and market-driven forces.

In contemporary Canada, for example, critics of a compromised public sphere decry the fact that new immigrants do not have equal access (for example, a medical doctor licensed in India would have to jump through too many bureaucratic hoops to be relicensed in Canada), and that the interests of First Nations peoples are routinely undermined. There is also the way branding and commodification imbues the public sphere with a logic that necessarily undermines its noninstrumental ideals. We must still buy something in Starbucks and submit to its overroasted aesthetic in order to sit on its patio and debate with our friends and neighbors. This important critique, however, does not go far enough in recognizing that the public sphere is also compromised by the shifting realities of globalization processes, shifts that expose how the ideals and assumptions of the public sphere are tied to the territory of the sovereign nation-state.[5] Incorporated into this understanding of the public sphere should be a recognition of how the public sphere is driven by a certain nation-centric assumption that does not account for the various ways the nation-state is not as stable and sovereign as it once was.

One way to elaborate this point is to turn to the immigration debate that stole the headlines in the United States during the spring of 2006. President Bush, while meeting in Mexico with Vicente Fox and newly elected Canadian prime minister Stephen Harper, staked out his position on the issue. Bush proposed that the nearly 12 million undocumented workers then living in the United States should be granted legal status and be allowed to remain temporarily on work visas. This was a position Bush already hinted at during the presidential campaign in 2004 when trying to rally support from the growing Hispanic constituency. The United States, Bush argued, needed an immigration system that "serves the American economy and reflects the American Dream."[6] But this simple phrase is not as clear as it seems. By recognizing that the U.S. economy depends on cheap and vulnerable labor is quite an admission, one that certainly set Bush apart from those on the far right of the Republican party who believed that any "wiggle room" given to undocumented workers is tantamount to giving hardened criminals a get-out-of-jail-free card.

Under Bush's proposal, illegal immigrants who had lived in the United States for five years or more, about 7 million people, would eventually be granted citizenship if they remained employed, had background checks, paid fines and back taxes, and learned English. Illegal immigrants who had lived in the United States for two to five years, about 3 million people, would have to travel to a United States border crossing and apply for temporary work visas. They would be eligible for permanent residency and citizenship over time, but they would have to wait several years longer for it. And those who had lived in the United States for less than two years, about 1 million people, would be required to leave the country altogether. They could apply for spots in the temporary worker program, but would not be guaranteed positions. On top of all of this, Bush repeated the old saw of tightening up border patrols, of placing hundreds of thousands of army reservists on the southern border from California to Texas.

One skeptical response is that this proposal to grant so many ad hoc green cards would encourage more undocumented workers to risk coming to the United States, thus reproducing the same shadow labor force five years down the road. If one shares this skepticism, then one is faced with the following question: If deportation of all the undocumented workers is unreasonable (not only because of the huge costs and the sheer impossibility of such a task, but because the U.S. economy is dependent on their labor) and if the United States cannot simply grant them all green cards, then what is to be done? To pose the question in this way forces us into the center of the global abyss.

When Vicente Fox was running for president of Mexico back in the spring of 2000, he spoke about the immigration problem between his country and the United States. In order to considerably reduce the illegal border crossings, Fox reasoned, there would first need to be some relative equality between the two countries' economies. Maybe on the level of three or four to one, Fox suggested. Only after such equity was met would Mexican workers stop putting their lives in the hands of unscrupulous coyotes, not to mention trigger-happy minutemen. Fox's proposal at the time was to move to a second phase of NAFTA, in which in five to ten years the borders between Mexico, the United States, and Canada would be open to the free flow of workers, the same free flow the first phase of NAFTA had established for products, services, and merchandise. Despite winning the election, Fox's open border proposal was lampooned as so much wishful thinking by many in the United States, and by the rich in his own country. It is one thing to talk about green cards and immigration policy, but to invoke the larger political economic system was perceived as ridiculous, if not downright blasphemous.

What is clear is that the economic inclusion of undocumented workers in North America depends upon political exclusion, in terms of benefits, status, and, perhaps most important, the right to participate in the political process. What was remarkable about the huge demonstrations in support of immigration reform throughout the United States in the spring of 2006 was that the undocumented workers themselves were the ones talking about their rights, not their universal human rights, but their rights as Americans. They reasoned that they were entitled to the same rights as all those citizens who require their cheap labor in order to live in a prosperous and relatively free society. It was this position, non-Americans demanding the rights of Americans, that enraged so many opposed to the demonstrators. Many Americans, appalled over the unabashed entitlement on display, asked incredulously: Who are these people to speak to us in such a way? The irony is that these critics were the same people who were aghast at so many Mexican flags waving at the first demonstrations earlier in the year. The leaders of the demonstration, acutely aware of the symbolic value of their media spectacle, asked those participating in subsequent demonstrations to switch to American flags. In the end, it was this switch (from defining oneself as a Mexican citizen who is no longer willing to remain in the shadows of the American economy and social system to arguing point-blank that one deserves the same rights as Americans and that in some time-crossed way one is already American by the mere fact that one lives and works in the United States) that stretched and challenged such key modern categories as citizenship, sovereignty, and national identity. Instead of citizenship that came from being born or naturalized in a certain national space, processes conferring the right to work legally and attain proper social rights, the provocation was that simply working in a country should confer rights that others in that country enjoy.

What could not be articulated in the debate was the possibility for a different kind of citizenship or permanent residency, one not based on any one nation, but a sort of transnational citizenship that confers rights based on something other than national identity. Indeed, there is a limit to even imagining such a policy, though many people are actually living these transnational identities every day. Not just the migrant workers, but more than 700,000 retired Canadians who spend six months of the year in

Florida or Palm Springs have a transnational identity. They own property in the United States, pay taxes, but return to Canada in order to retain their health benefits, for in Ontario a Canadian forgoes access to public health services if they live outside of the province for more than six months a year. Should these snowbirds be granted green cards as well?

During the demonstrations, Canadian officials remained quiet and there was a conspicuous lack of national reporting about immigration. There were many in Canada who were in the same boat as Hermelindo Gutièrrez and could certainly benefit from Canadian status, but in Canada most were concerned with keeping the border relatively open to American goods and tourists while not appearing to be soft on the dangers of global terrorism.

One of the key points here is that national political problems are global economic problems. One cannot approach the one set of problems without engaging the other. At the same time, global political problems are national economic problems. And there is no better example of this than the threat of global terrorism. This brings us to another phenomenon widening the entrance to the first global abyss: security following 9/11.

For the past century, U.S. immigration policy looked primarily to Mexico. In fact, when George Bush met with Mexican president Fox just before 9/11 in order to prioritize migration issues and the southern border, he canceled a similar meeting with Canadian immigration officials.[7] By the end of 2002, a radical shift occurred in which Canada took precedence over Mexico, primarily to manage American fears about Canada being used as a base for terrorists. A harmonizing of policy and a common security perimeter were discussed. The new policy included tighter control at border crossings (including racial profiling of Canadian citizens trying to enter the United States) and significant changes to the process of granting immigrant visas. A majority of Americans believed that improper enforcement of immigration laws at the border had been responsible for the 9/11 terrorist infiltration, even though all nineteen of the terrorists entered the United States legally as visa students and not as immigrants or refugee claimants. The inextricability of migration and security (both in terms of discourse and policy) was complete, with the privacy of information, refugee claimants, and immigrant rights all taking a back seat to new security concerns.

One complication is that those enemies trying to sneak into a country from the outside are not necessarily the only ones who wish to commit terrorist acts. This proved to be the case back in 1995 when Timothy

McVeigh detonated bombs in front of the Oklahoma City federal building. Or we could point to the four British-born suicide bombers who killed over fifty innocent commuters on a London bus in the summer of 2005. And there is the Canadian example: in the summer of 2006, Toronto police arrested seventeen Canadian citizens as part of an undercover sting operation alleging they were plotting a local terrorist attack with three times the explosive material used in the Oklahoma bombing. This "homegrown terror" by second-generation Canadians who "succumbed to Fundamentalism," as the Canadian major newspapers put it before the facts of the case were disclosed, was a severe challenge to older paradigms of inside and outside, us and them, and national and global. In fact, the number of times the suspects were labeled as "Canadian-born" citizens as opposed to simply "Canadian" clearly indicates an emerging two-tier structure of citizenship that exists even in a country as multicultural as Canada. These changes testify to a slow but significant shift away from powerful nation-based identities that have characterized the modern era.

Today, many of us are tired of hearing about globalization and receive it with a healthy dose of skepticism and disinterest. Like any new god-term, globalization is employed as a certain destination, the end of something, the limit whose other side cannot be imagined. This is not to imply that the world has actually achieved complete globalization or that the name and the desire for this condition has anything to do with its attainment, rather that the hype of globalization has come and gone. When the xenophobes, anarchists, business leaders, neonationalists, fundamentalists, academics, and Luddites all speak the same language (as they seem now to do when dutifully reciting globalization's mantra, in either its utopian or dystopian dispensation) you know that something's over, that something's been relegated to the tedium of trendy dissertation topics and best-selling nonfiction.

Globalization is, therefore, a struggle over how to tell the story of globalization. Is globalization a new period in human history in which time and space function differently, in which all the old categories of economics and sociology and literature no longer work? Or does globalization mark a continuation of a process begun with the Industrial Revolution and perfectly anticipated by so many prescient nineteenth-century thinkers from Smith to Dickens? The only way to answer this is to try to set some parameters around the processes of globalization itself. As a category to explain a certain political-economic and cultural logic, globalization does represent a qualitatively significant shift, something that comes into being over the past thirty or forty years. This is not simply hype, for there is something very real about the new information technologies and the flexibility of commodity production, about the growing significance of speculation and currency markets, about the corporate consolidation of food, medicine, weapons, and entertainment as the battle over intellectual property (located in a global, conceptual space rather than in a national, physical space) emerges as one of the great geopolitical crucibles of the day. Globalization is not a different mode of production than capitalism, but a different stage of capitalist development—one that shapes the singularity of the global abyss. For now, however, it is crucial to recognize how nations, national policies, and nationalisms still matter, to the extent that they matter within the qualitatively different space of globalization. And a perfect example of this is how Canadian immigration policy flies in the face of the global labor market's most vital requirement of expansion.

National Sickness and Global Health

I was also between two different national health care systems, one in the United States and the other in Canada, that were both integrated into an emerging globalized system. The global abyss was produced here by the noncoincidence of the nation-state and citizenship model of health care administration and the emerging global citizen (or noncitizen) who could not be accommodated by any current healthcare system. We know that the U.S. system is organized around private insurance companies (HMOs and the like) with many linked to employers, while the Canadian system is organized around public institutions linked to immigration status. When I was employed by a U.S. university, my health care was wholly provided by a private university policy. In Canada, my health care is provided by the provincial health care plan (with extended health care, including prescription drugs and other services, provided through the private university policy). It is for this reason that health care costs would not be as critical to U.S. immigration decisions, since the states and federal government would not be responsible in the same way as the Canadian public purse. It makes perfect sense that different nations have different health care systems, but what complicates matters, and what opens up this second global abyss, is that such national differences exist together with global consolidations in medical research, law, protocols, disease, technologies, personnel, patients, and the production and distribution of pharmaceuticals.

Of the approximately 6,000 total scientists working at the National Institutes of Health in 2005, approximately 2,600 were foreign postdoctoral fellows and other visiting scientists.[8] In the 1990s alone, the number of non-U.S. clinical investigators conducting drug research for the Food and Drug Administration increased sixteen-fold.[9] And then there is the outsourcing of clinical trials, in which India has emerged as a crucial player. Since costs for U.S. drug trials regularly reach $800 million, all the leading pharmaceutical corporations exploit the Indian market, where costs are usually 60 percent lower.[10] With a quarter of clinical trials conducted in developing countries not undergoing ethical review, the likelihood of corruption is high. In India in 2003, more than 400 women who could not conceive were enrolled in a trial without their knowledge or consent, to see if a drug called letrozole induced ovulation.[11]

These global flows of medical research follow many of the same routes as the drugs they help discover. Not only is production outsourced all over the world like any other commodity today, but distributors have also gotten in on the act. In 2005, Canadian online pharmacies, for example, provided about $800 million worth of low-cost drugs a year to 2 million uninsured and underinsured Americans.[12] These Internet pharmacies, mostly with warehouses located in rural Canada, were buying large quantities of drugs from the major pharmaceutical corporations and then underselling these very corporations abroad under the banner of free trade. A loophole such as this is usually closed by the industry heavyweights, but in this case it was hard for those who usually bring down the hammer to resist the sympathetic images of busloads of elderly Americans crossing the Canadian border in order to buy cheap drugs, together with an American health care system whose costs are so high that almost two-thirds of bankruptcies are due to the inability to pay for basic health care.[13]

But it is not only drugs and researchers that are crossing borders; patients too (at least the ones who can afford to) are traveling the globe, searching for new drug trials and experimental therapies. Most leading cancer hospitals in the United States now advertise services and solicit foreign patients, while European spas, such as the classic spa in Baden-Baden, still attract the global rich who are in search for the ideal place to convalesce.[14] The supplement to this medical tourism is a growing trade in human organs, a trade in which poor Brazilians are flown to South Africa to sell their kidneys to desperate Israelis. This black market also thrives in India, China, and the Philippines, not to mention in Britain, where Stephen Frears's film *Dirty Pretty Things* (2002) poignantly highlights how such rackets are related to immigration and fears of undocumented workers.

On one level, there might be nothing new about such medical traffic. Border crossings in search of cures and medical knowledge go as far back to the first expeditions of discovery. But there is something qualitatively different about our current situation. For example, the global AIDS pandemic is not only the most perfect metaphor for globalization; it is globalization. Money, goods, people, information, and disease travel more flexibly and at speeds hitherto unimaginable. Despite the integration of the planet, the sub-Saharan African HIV population is left to die, while in the developed world infection rates have leveled off and survival rates have increased. The global AIDS pandemic is not only about uneducated Africans who primitively ascribe their illnesses to local mythology; nor is it only about unscrupulous African officials who refuse to spend money on antiretroviral medicine. It is also, and more profoundly, about transformations in the world system: transnational institutions (pharmaceutical corporations and the WTO) that are not at liberty to suspend the logic of competition and profit; new global flows in the sex industry; new labor flows that are mobile and gendered; and a post–Cold War world order in which the political motivation for supporting the global south is at an all-time low.

We must remember that in the days of decolonization, the 1950s and 1960s, public health was a priority in most African nations, and infant mortality and life-expectancy rates improved dramatically. After these countries experienced economic crises in the 1980s, health care budgets were slashed, and the World Bank and the IMF imposed austerity measures known as structural adjustment programs (SAPs). These programs, imposed on more than seventy developing counties, were the fulcrums on which globalization turned. The SAPs were destined to promote exports made cheap by currency devaluation and reduce "inefficient" government spending to allow nations to earn enough foreign exchange to pay back their loans. But the benefits from the new programs were never properly distributed to the nations as a whole. The measures also created conditions conducive to the spread of HIV, by displacing young women and children from rural villages to cities, where they resorted to commercial sex work, and by displacing groups of men to urban areas, where they were more likely to engage in unsafe sexual practices. As health care spending dropped, AIDS education and treatment programs evaporated, and the conditions for an AIDS crisis were optimized. The HIV and AIDS crisis is as much a symptom of globalization as it is of immune deficiency.

The same can be said about mental illness in Japan—that it is as much a symptom of globalization as it is of depression. In the late 1980s, Eli Lilly, the maker of Prozac, decided against selling in Japan because of the cultural argument that the Japanese inability to talk about "shameful" conditions, combined with a medical establishment in which patients rarely speak openly to doctors (and vice versa), would make it very hard to cultivate a profitable market for antidepressants.[15] Today, the use of antidepressants is growing at a rate unmatched by any other class of drugs in the history of Japanese medicine. When the crown prince expressed publicly in August 2004 that his wife, Crown Princess Masako, was suffering from depression and currently taking an SSRI, no one was that surprised, given that between 1998 and 2003, sales of antidepressants in Japan quintupled, while GlaxoSmithKline saw its sales of Paxil increase from $108 million in 2001 to $298 million in 2003. According to the company, during one seven-month advertising campaign, 110,000 people, in a population of 127 million, consulted their doctors about depression.[16] Trying to explain such a phenomenon is always tricky, and always at risk of employing the most vulgar stereotypes of a foreign culture. Trotted out in this case are the usual Orientalist tropes, such as that the Japanese are more morose and, therefore, more prone to depression, or that the Japanese mindlessly follow the latest boom, from electronics to psychotropics.

The other side of this cultural argument is a certain materialist argument that shapes the public debate about antidepressants in Japan, as it has in many Western countries. For example, those on the anticonsumerist left began by arguing that the big pharmaceutical corporations are producing and shaping the very category of depression in order to offer the cure in the form of a hugely profitable commodity. They then inverted this position by accepting depression as the illness par excellence of our late capitalist moment of alienation, fragmentation, and vapid consumerism. It is no surprise we need such drugs, so the argument goes, given the deadening environments in which we exist.[17]

Both AIDS and depression, therefore, are rather special to our current historical moment. We might even call them global illnesses. Of course, depression has existed long before current discussions about globalization, but the point is that the current dominant form of depression functions differently today than it did in the past, in terms of how its ubiquity and focus in popular culture is inextricably linked to the rise in transnational pharmaceutical corporations as well as to late capitalist societies. We can think of cancer in the same way. The World Health Organization (WHO), which tracks changes in cancer for developing and industrialized countries, and which distinguishes differences in rates due to detection technologies and life practices, has recently noted a growing global similarity in rates and types of cancer. The WHO estimates that by 2020 cancer rates could increase by 50 percent, with prior differences giving way to significant similarities. More important, the global differences in imagining cancer, from its mythical and metaphorical meanings to its causes and effects, are evening out.

Despite this global convergence of disease, however, health care policy and access to drugs differ radically, depending on the nation in which a patient lives.[18] And this is precisely where the global abyss forms: in between the current configuration of disease (its research, care, and therapies), which turns on globalization processes, and the access to health care that is determined by national citizenship (and, now more than ever, class standing within nations). This abyss is abundantly apparent when one looks at some of the new Internet patient communities. In my case, I joined a chronic myelogenous leukemia (CML) group shortly after diagnosis. The group was administered in the United States, but consisted of members from over thirty countries. The group formed right when one of the most important advancements in the treatment of CML was under way. Before 2001, the first-line treatment for CML was either interferon, a self-administered injection that seemed to suppress the acceleration of the disease, extending the average lifespan from about four to six years (but killing off healthy as well as cancerous cells) or a bone marrow transplant (for which not everyone is matched and to which high mortality and morbidity rates are attached). Since CML is a fairly easy disease to understand (in terms of its chromosomal mutation and genetic makeup), many new experiments in targeted therapy turned to CML despite its having a relatively small population.

When the Internet group learned about one of these promising new experiments, news spread quickly among researchers, clinicians, and pharmaceutical executives, who found themselves on the receiving end of a rather desperate and well-connected patient community. Novartis, the sixth-largest pharmaceutical corporation in the world at the time, was behind the medical trials and the manufacturing of the drug Gleevec, but was not expecting to prioritize production and further development (mostly due to the relatively small patient population of the disease). Nevertheless, Gleevec gained approval from the FDA (and from Health Canada and a number of other national health care systems) faster than any drug in the history of modern medicine. Once the drug was approved and showing breathtaking results, the CML community the world over came knocking on Novartis's door. After a concerted Internet petition drive that was global in nature, Novartis felt pressured to crank up its efforts to bring Gleevec to market—making it the second-most profitable drug for the Swiss company.

After I was diagnosed, Gleevec was approved as first-line treatment in the United States, but not yet in Canada (where patients first had to fail interferon in order to be prescribed the Novartis drug). In the United Kingdom, approval for first-line treatment of Gleevec took much longer than in other countries precisely because the National Institute for Clinical Excellence was not convinced that Gleevec was that much of an improvement on interferon. These national differences were not just about different drug-approval protocols; they were shaped more notably by Gleevec's enormous costs. Novartis priced the recommended dosage of Gleevec at the same price as interferon, which costs approximately $45,000 a year— an amount that was the same the world over under a global pricing scheme (including exclusive licensing agreements), a pricing scheme that is indispensable for big pharmaceutical corporations not wanting to be undersold by black-market distributors exploiting fluctuating exchange rates. Before Gleevec was approved in various countries, Novartis enrolled over 7,500 patients in clinical trials that took place in several countries. After approval, patients were then left at the mercy of their respective national health care systems. And this is when the abyss opened.

Many on the Internet group (sponsored by Yahoo and run out of the United States) were faced with a situation in which the force of the collective, based on our common health, was now punctured by national differences. The group began to wither and grow more confrontational. How could we act as a community when separated by such fatal inequalities? Our Japanese members were getting Gleevec provided at 100 percent of cost, while those from South Korea had to pay 30 percent. In New Zealand there was a decision not to put Gleevec on the government list of approved drugs, primarily because the number of people with CML (ninety in all) did not justify such costs. And then there were patients from all over the global south who were desperately writing to the Internet group asking for any help (perhaps people could double-up on their dosage, have their insurance pay for the whole prescription and then send half to those who would die if they went without). Allegiances were tested and some strange bedfellows were made. Novartis, for instance, financed and trained local patient groups on how to forcefully lobby their respective governments— each government that put Gleevec on its list of approved drugs meant millions in profit. The Internet group, composed of well-intentioned, supportive, and frightened patients and caregivers, was rendered powerless and then found members at each other's throats, due to rock-bottom political-economic realities.

The disagreements, however, were actually less about national differences than political differences. One member would write about how the Bush administration's privatization policies were disastrous for cancer patients, while someone would respond that socialized medicine was not the answer. When tempers would flare, the invariable plea would come to stop all politics and act solely as a support and information group. This would then be followed by an appeal not to surrender the power of the group to shape both corporate and public policy, especially since the group had so much success influencing Novartis to bring the drug to market in the first place. One theme developed around how many in the developing world had little access to Gleevec and how Novartis's newly formed patient assistance program was failing to close the gap. Invariably, however, the conversation would come back to U.S. politics, which was partly due to the group's being dominated by Americans and conducted in English, but was also because many of the American members had no other way to under-stand power and the high-stakes issues (health care, corporations, geopoli-tics) than through a rigid narrative of Democrat-Republican plot points.

The group split, with a certain wing forming an Asian CML group that included Americans. When push came to shove, the powerful identity of national difference trumped the global identity of cancer patient—espe-cially when the stakes were as high as who would receive a life-saving medication and who would not. There is a lesson here for global resistance politics today: despite the fact that the leaders of supranational institu-tions (transnational corporations, the IMF, and others) hold more power than most national politicians around the world, a global resistance will need to respect how older national configurations still shape the rules of the game.

Another aspect of this second global abyss of national health care sys-tems within a larger globalizing system is how private corporations are cashing in on the public resources of individual nations. Whenever Novar-tis's executives are questioned about the high price of Gleevec, its leaders bring out the old saw of research and development. How could a company possibly recoup its investments, so the argument goes, if its products were priced for accessibility rather than for fair compensation? The bottom line for pharmaceutical executives is that if there were not exorbitant pricing then there would not be such magnificent drugs. But recent studies have shown that the largest proportion of the pharmaceutical corporations' revenues is directed toward marketing, advertising, and administration (at 27 percent), followed by profits (at 18 percent), with money spent on research and development only making up 11 percent of revenues.[19] Even more significant is that most of the research that leads to the most success-ful medications come out of publicly funded institutions.[20] As for Gleevec, the basic research on CML over the past forty years and the initial clinical investigations for STI 571 (later named Gleevec) came out of U.S. university laboratories. This partnership between the private and public sectors be-came easier to build following the Bayh-Dole Act of 1980, which gave academic scientists an automatic right to patent and sell their federally funded work to private corporations.

This was going on at the time when many in the Congress were worried about an innovation crisis, about the brightest scientists leaving univer-sities unless universities could match the salaries and bonuses of the private sector. The breakdown of private and public became even more acute in the 1990s, when partnerships with business became the watchwords of the contemporary university. For example, in 1995 Novartis embarked on a five-year $25 million investment in the Department of Plant and Microbiology at the University of California, Berkeley, an investment amounting to almost one-third of the department's budget. This corporate involvement occurred when California decreased its subsidizing budget for public universities. Taxpayers end up financing national institutions that produce knowledge to be commodified by global corporations who are only held responsible by shareholders (not by the taxpayers who helped finance the projects in the first place). Moreover, Gleevec was granted "orphan drug" status by the FDA, thus allowing Novartis to benefit from federal tax rebates equal to one-half the cost of clinical testing.[21]

To make matters worse, when the nation tries to intervene on behalf of its citizens, the pharmaceutical corporations punish the citizens. For example, when the Indian company Natco Pharmaceuticals received government approval to make a generic Gleevec (named Veenet) that would cost $377 a month, Novartis immediately refused to approve applications by Indian nationals hoping to participate in the company's discount drug program. But as we learned from the HIV situation, even the most unapologetic neoclassical economists find it hard to brandish so many economic theories in the face of dying people. Always one step ahead of the economists, Novartis went ahead to purchase some of the world's leading generics companies, effectively updating the old adage of how to get along with a competitor to "if you can't beat them, buy them." Things are global when they benefit the corporations, and national when they do not; likewise things are national when they benefit the national leaders, and global when they do not. Unlike the way the globalization debate is organized around the zero-sum logic in which some argue that the nation is as strong as ever while others insist that the global has displaced the national as the key unit of power, the CML case exposes how this debate is utterly flawed and that the relationship between the national and the global is much more complex. The new mutuality of nations and global institutions, rather, produces spaces and experiences that remain unthinkable to those whose business it is to examine such matters.

Over the past few years, the pharmaceutical industry has hinted at a new justification for its pricing practices. In addition to making the argument that high prices are required in order to recoup research and development costs, some pharmaceutical executives are now arguing that prices can go through the roof simply because the medical results from their products are so astonishing and the value of extending life has no limit. They ask, discounting all economic and actuarial reality, who would say that a life is not worth forty, sixty, or one hundred thousand dollars a year? When Genentech, for example, indicated it would effectively double the annual price of its colon cancer drug, Avastin, to about $100,000, a drug also used to treat breast cancer and lung cancer patients, the company's executives argued that the increase was based on "the value of innovation, and the value of new therapies."[22] Given that Avastin was projected to earn nearly $7 billion in the United States alone by 2009, the justification certainly had to shift from a quantitative one based on research and development reimbursements to a qualitative one based on the pricelessness of life.

With such new justifications for the exorbitant costs of new cancer drugs, it is no wonder governments and insurance companies are concerned. In fact, in June 2006, the Canadian Patented Medicine Prices Review Board (which regulates the price of patented medicines to ensure they are not excessive) determined that the colorectal cancer drug Erbitux was simply too expensive. As a result, Bristol-Myers Squibb, the maker of Erbitux, refused to launch the drug in Canada, even though it had been approved by Health Canada. "For the time being, we made the decision not to launch Erbitux because the value is not in line with the innovation it brings to patients," said a company spokesperson.[23] The value, estimated to be about $27,000 for a four-month supply of therapy, is not in line with the innovation, an innovation that does not yet prove to extend the lives of patients, but rather delays tumor growth. But if Erbitux was the silver bullet, would a million dollars a year still be "not in line with the innovation"?

If the Canadian government had to pay for my Gleevec, I would certainly present excessive demand. And this is why having private insurance was so integral to my appeal of the government's rejection of my immigration application. But as the case progressed, the government refined its argument against me: although I did have my drugs paid for by private insurance, they now reasoned, I could quit my job, therefore requiring public assistance for which the government might be responsible. In Ontario the Trillium Program offers assistance to uninsured and low-income patients who cannot pay for prescription drugs on their own. Gleevec is not one of the drugs covered by Trillium, but there is the possibility of appealing to the Trillium program to include drugs that are not normally covered. The government's argument, therefore, went something like this: I might get fired (even though there has not been a single case of a tenured professor being fired at the University of Toronto in the past eighty years) or quit my job (quitting without having another job that would offer comparable health care or having another job that would pay so little that I would fall within Trillium's indigent category), and then I would apply to the Trillium Program, get rejected because it does not cover Gleevec, then appeal that decision and request that Gleevec be considered an exception, and then finally win the appeal which would mean that, in the end, Canadian taxpayers would pay my exorbitant health care costs.

The likelihood of such a scenario is rather far-fetched, or to use the legal language, "not reasonably expected to occur"—which is, indeed, the key criterion by which the law is judged. In fact, my original rejection was based on the government's judgment that it is within reason that over a five-year period I would stress the health care system more than the average Canadian. In other words, it is not within reason that an applicant without any present medical condition will be struck by lightning and need expensive life-saving medical procedures. Thus, even though any new immigrant can immediately become a hefty financial burden on the government, exceptional scenarios cannot be considered grounds for inadmissibility. Why, then, would the government prepare an argument against me based on such an exceptional scenario? Or to put this another way: Why is the government allowed to be so unreasonable? Reason is always defined by the state, by those in power, and, therefore, those in power assume that the state is always reasonable—the perfect circle. Of course, reason here has no content, and what fills its form can change instrumentally from one moment to the next. And the really sad part is that this assumption informs the larger debate over health care on a global scale—that somehow it is reasonable that the profit motive will determine who lives and who dies.

The stakes of my case were quite high regarding the future of Canadian immigration and health care policy. At the time of my appeal, the supreme court of Canada was deciding a similar case in which two applicants for permanent residency were denied entry based on the health conditions of their children. David Hilewitz, from South Africa, and Dirk de Jong, from the Netherlands, both applied for permanent residence in their names and in that of their families. Both qualified on their own, but were denied admission due to the intellectual disability of a dependent child who "would cause or might reasonably be expected to cause excessive demands on health or social services" in Canada.

David Hilewitz's case applied for permanent residence in 1999 under the "investor" category, which requires an applicant to have substantial business experience and a net worth of at least $800,000. The applicant must also commit to making a significant financial investment in Canada. Hilewitz satisfied these requirements, but it was his seventeen-year-old son who complicated matters and led to the denial of David's petition. His son was born with intellectual disabilities and was, therefore, deemed a prospective burden on social services, such as special schools, vocational training, and respite care. Hilewitz pointed out that his son had never used publicly funded schooling in South Africa and that the family had helped establish a special school for him and others with similar disabilities. Since his son was interested in computers, moreover, Hilewitz offered to establish a computer gaming business so that his son would always have secured employment. Hilewitz wrote, "We have never been a drain on any institutional or social service structure to support our son and cannot conceivably ever contemplate any change to this ethos in the future."[24] The legal question was whether the financial resources that otherwise qualified Hilewitz for admission to Canada could be disregarded in assessing the demand on Canada's social services of his child's disability.

One way to get at this is to quickly trace the history of excessive demand throughout Canadian immigration law. The first statute was written in 1859 and dealt with the admission of persons with physical or mental disabilities who might impose financial burdens on the state. Under An Act Respecting Emigrants and Quarantine, a medical superintendent would board ships arriving at Canada to check for "any Lunatic, Idiotic, Deaf and Dumb, Blind or Infirm Person, not belonging to any Emigrant family, [if] such person is, in the opinion of the Medical Superintendent, likely to become permanently a public charge."[25] If an emigrant family could care for the disabled person, entry would be permitted. An exception was written into the act of 1906 as follows: "Unless he [the disabled] belongs to a family accompanying him or already in Canada and which gives security, satisfactory to the Minister, and in conformity with the regulations in that behalf, if any, for his permanent support if admitted into Canada."[26]

This changed with the Immigration Act of 1910, which distinguished between those with mental and physical disabilities, instituting an absolute prohibition on admission for those with mental disabilities. Individuals who were "physically defective," however, could be admitted with evidence of earning capacity or family support. In 1919, this prohibition changed to either the "mentally *or* physically defective to such a degree as to affect their ability to earn a living."[27] Thus, it no longer mattered if there was sufficient family support; rather, a burden threshold was applied equally to all—a threshold that remained in the law until 1952.

The Immigration Act of 1976 replaced the absolute rejection of "prohibited classes" with "excessive demands," in order to account for individual cases. And this brings us to the act of 1985 and the insertion of reasonableness into the language of the policy. This act reads: "No person shall be granted admission who is a member of any of the following classes: (*a*) persons who are suffering from any disease, disorder, disability or other health impairment as a result of the nature, severity or probable duration of which, in the opinion of a medical officer concurred in by at least one other medical officer, (ii) their admission would cause or might reasonably be expected to cause excessive demands on health or social services." In the most recent Immigration Act of 2002, the phrase "would cause" was dropped, leaving only "might reasonably be expected to cause excessive demands."[28]

When cross-examined by Hilewitz's lawyer, the visa officer who originally denied the application explained that "if Mr. Hilewitz, Heaven forbid, were to become insolvent, it is reasonable to assume that Gavin would use existing public services." In other words, the visa officer considered various contingencies and based her rejection of Hilewitz on such a future possibility. In a seven-to-two decision, the supreme court overturned the denial, arguing, "The threshold is reasonable probability, not remote possibility. It should be more likely than not, based on a family's circumstances, that the contingencies will materialize."[29] Justice Rosalie Abella went on to write, "It follows from the preceding analysis that the Hilewitz and de Jong families' ability and willingness to attenuate the burden on the public purse that would otherwise be created by their intellectually disabled children are relevant factors in determining whether those children might reasonably be expected to cause excessive demands on Canada's social services."[30]

Nonetheless, these cases are significantly different from my own. First, my case is directed toward health services, not social services. Second, it is unclear whether or not someone in remission from leukemia should be considered disabled. Third, I was diagnosed with leukemia after already living and working in Canada for two years. Fourth, while in remission, I will not require any special health or social services from the government.

The similarity to my own case has to do with an individual assessment in which a flat prohibition is unfair. In both cases a remote speculation of worst-case scenarios must be held to some reasonable standard. Most important, what is shared is a medical-economic model of immigration in which the costs and benefits of a prospective immigrant are based on how much they contribute to the Canadian economy in terms of their labor power. Since everyone will die and burden the health care system, the most important variable according to this model is how economically productive one can be in the meantime. And those with mental and physical disabilities are usually judged as failing to contribute in such specific ways. When we look at the most heinous forms of discrimination throughout the history of Canadian immigration, we witness a similar economic imperative at work. From head taxes on Chinese to assuming that Jews could not live outside metropolitan centers, from discouraging black farmers to emigrate based on their "inability to acclimate to the harsh Canadian winter" to the "continuous passage" policies requiring immigrants to travel directly from their home country, thus closing the door to many Japanese and South Asians, such policies almost always employed an economic justification to cover their discriminatory core. Today these blatant acts of discrimination are viewed as regrettable and archaic forms of the past, far away from present sensibilities and policies. But what links this past with the present is a reliance on a very narrow definition of productivity—one that totally dismisses a social model that values the contributions of the disabled or the ill on an entirely different scale, one that might emphasize, for example, how an autistic boy inspires his friends to view the world in rich, alternative ways.

National Cures and Global Management

I am between a transforming cancer industry in which the old categories of cure and remission no longer hold as much clinical purchase, though they still structure the social aspects of the illness. The global abyss, in this case, is widened by the time lag between rapidly shifting categories of cure and management and a residual and slow-moving health care system working with outdated concepts. In this time lag, a space opens.

For example, when I took my medical examination as part of the application for permanent residency, the doctor had to check one of four boxes regarding my health. The first box signaled findings that are unremarkable or minor conditions that required no exceptional care. The second box signaled findings that required periodic specialist follow-up care, but which normally can be handled without resorting to repeated hospitalizations. In this category the applicant should be able to function independently and be self-sufficient; for example, the applicant might suffer from asymptomatic congenital or rheumatic heart disease where the demand for hospitalization or surgery appears very unlikely over the next five to ten years. The third category signaled findings that may require more extensive investigations or care, such as dementia, Parkinson's disease, multiple sclerosis, or symptomatic heart disease. This third category is reserved for those requiring home care, major hospitalization, specialized hospital facilities, or for those in which deterioration appears quite likely and self-sufficiency appears doubtful (for example, when active tuberculosis appears to be present or behavior appears to be potentially dangerous to others, such as might occur with illicit drug use or alcohol abuse). The fourth box was for other conditions. Since my blood test came back normal (as expected, since I am in remission), the immigration-appointed physician was prepared to check the unremarkable box. He then asked about my medical history and I disclosed my leukemia diagnosis. After reading a letter from my specialist about my remission and the success of Gleevec therapy, the doctor ended up checking the second box. But the fact that the categories are so rigidly defined is symptomatic of the current unevenness between medical paradigm shifts and the cultural and social ways we register illness.

This no doubt partly explains why the refusal letter I received from Immigration Canada was so misrepresentative of my case. In my application, my doctors took great pains to explain that I was taking Gleevec, an oral medication that is not publicly funded by OHIP. The refusal letter then stated that I was undergoing chemotherapy, which is a very expensive public therapy. This misinformation was so blatant that it seemed more like a failure of imagination than a careless misreading of documents.

In other words, the designated immigration official simply could not imagine that leukemia would not require costly public services, most notably chemotherapy. The enduring equation "cancer equals death" was doubtless behind this failure.

Susan Sontag's classic essay *Illness as Metaphor* (1978) exposed the damage done by the metaphoric use of cancer, metaphors (economic, political, and cultural) that always imply an inescapable fatality, a hopeless nothing-to-be-done disgust. Such shameful disease metaphors, Sontag argued, were used in the late 1970s to reflect and excuse a capitalist society not out of balance, but repressive and unable to properly regulate consumption and violence. Sontag ends her polemic with the following perceptive sentence: "The cancer metaphor will be made obsolete, I would predict, long before the problems it has reflected so vividly will be resolved."[31] Ten years later, Sontag quite astutely turned to AIDS and its metaphors to show where the earlier ideological work of cancer metaphors was now being performed. Sontag's point that cancer metaphors served to avoid a straight-up examination of the world's problems should be rethought today. In fact, due to emerging medical technologies and the celebrated genetics revolution, cancer has returned to offer up all new hope for engaging the problems of the world. As I argued previously, oncologists speak more about disease management than cure, and the category of remission is losing its value.

Things will really get messy when medical preemption becomes more common. With detection technologies revealing latent illnesses, it is no surprise that so much financial and emotional investment is placed in stem cell research, especially procedures in which a patient's own stem cells (those not terminally differentiated) are extracted and then inserted into ailing organs in order to regenerate deteriorating cells. No doubt the term "preemption" has an all too familiar ring—but this is not another case of bad metaphoric usage. Rather, there is a shared logic in the way preemption was employed by the Bush administration to justify its attacks on Iraq and the way preemption is now employed in economics, psychiatry, ecology, culture, and the medical sciences. And this has everything to do with the current way crisis is used and abused in everyday life.

In the period before to the Second Gulf War, Iraq had invaded no country and had no more provoked others with nuclear or chemical weapons than North Korea, Israel, India, Pakistan, China, or the United States.

But the Bush logic was clear: Saddam Hussein was a time bomb, a "not if, but a when" threat that had to be snuffed out sooner rather than later. The U.S. argument for preemptive war (the Bush Doctrine) came down to one of self-defense under Article 51 of the United Nations Charter, a case of self-defense in which, according to the Bush administration, the traditional strategies of deterrence and containment were no longer sufficient. As Bush himself had argued back in June 2002, "Deterrence means nothing against shadowy terrorist networks with no nation or citizens to defend and containment can't work when unbalanced dictators with weapons of mass destruction can deliver those weapons on missiles or secretly provide them to terrorist allies." For Bush, the world had changed: "If we wait for threats to fully materialize, we will have waited too long." Bush invoked the changes in the world-system (in fact he invoked a breakdown in the nation-state-based world-system as well as in the U.N. itself) to legitimate his new preemptive military campaign (that would be carried out with or without the U.N.).

But does not the IMF argue the same thing? Rather than simply coming to the aid of crisis-ridden countries, the IMF now desires to preempt financial crises by prescribing the liberalization of capital markets and austere monetary and fiscal policies. Bail-ins, instead of bail-outs, purgatives instead of palliatives, early lending instead of lending too late, this is the mantra of the IMF, a mantra that has effectively served the interests of the financial markets and transnational corporations more than global economic stability or countries in crisis. In fact, it was precisely these policies (followed by a deadly prescription of sharply increased interest rates) that produced rather than preempted the East Asian financial crisis, not to mention the crises in Brazil and Argentina.

This returns us to preemptive medicine. Trying to fix the crisis of cancer or any other serious illness before the development of advancements in medical research and without simultaneously engaging the social structures in which such advancements exist only produces a crisis of another kind—a crisis that is sure to spread not only greater injustices and violence, but (and here is where the circle closes) greater threats to human health itself, especially in the form of mental illness and the biological maladies caused by environmental damage. For instance, there is no way under the present system that medical technologies will ever be properly democratized; in fact, it is precisely around the inequality of cutting-edge health care where we will see the impossibility of democracy under the current system of global capitalism.

But perhaps the present debate will shift all by itself, especially with the emerging prevalence of preemptive medicine. As defective hearts are repaired with stem cell procedures and future cancers are prevented by genetic engineering, the unequal realities of everyday life will simply become impossible to conceal. The harsh realities will become more transparent as the ideologies of democracy and equality weaken and become harder to sell. A global two-tier system will most likely grow even stronger than it is right now, a system in which only those who can afford to save their lives will be allowed to do so. To justify such a system, there will be the employment of a permanent state of crisis. Those in power will doubtless shift rhetoric, from one that apologizes for democracy's slipups to one that justifies democracy's suspension: "We would like for everyone to have access to such cutting-edge health care, but given the current crisis (war, environment, economy) we simply cannot afford to do so in the meantime." Here the problem is the extension of the meantime, an extension in which the meantime becomes the permanent destination, rather than a temporary moment of exception. Now the new chronic becomes not only a management strategy to treat the seriously ill patient, but the perfect justification to withhold care from the economically weak.

We often hear reports of people faking their illnesses or those of their loved ones. It seems that not a month goes by without another case of a mother who lied about her son's cancer as a way to raise money and elicit sympathy. A disorder known as the Munchausen syndrome, named after the eighteenth-century German officer Baron von Munchausen, who was known for embellishing the stories of his life and experiences, refers to a psychological disorder in which physical symptoms of a disease are feigned. There are many ways to account for such fraudulence, ranging from the desire for sympathy to the need for money. Indeed, the man who lies about his cancer is almost certainly suffering from serious mental health problems—perhaps as severe and dangerous as any cancer. At stake here is the difference between mental and physical illnesses. In the earlier review of Canadian immigration policy, we saw a clear prejudice against those immigrants with mental illnesses. In 1910, anyone identified with a mental condition was absolutely prohibited from entering the country, whereas leeway was granted to those with physical disabilities, depending on the stability of their emigrant family.

A few years ago the Canadian Psychiatric Research Foundation launched a campaign that confronted this problem. Through TV, radio, and print advertisements, some of the dirtiest assumptions about the mentally ill were revealed. In one of the radio spots, a woman calls 911 to report a seriously injured man who has just been hit by a car and is lying flat on the pavement. The caller is scared and frantic as she communicates the accident. The dispatcher, unconcerned and nonchalant, asks the caller if the man is bleeding. "No, he doesn't look to be," the caller confirms. "He's probably fine, then," the dispatcher says condescendingly. Stunned and confused, the caller adds, "But I just saw him get hit." The dispatcher responds, "Don't worry ma'am, he's probably just looking for attention . . . either that or he just doesn't want to go to work. Just walk away, he'll have to learn to deal with his problems like everyone else." At the end of the spot, a voice asks, "Imagine if we treated everyone like we treat the mentally ill?" In the print advertisements, this same question is repeated in smaller print below very large text that reads, "A lot of people get cancer because they just can't deal with reality."[32]

Whereas cancer remains an indelible reality for many no matter how total their remission, many people suffering from mental health disorders are forced to act out their symptoms so that others will take them seriously and see them as something other than a fraud. It is almost always harder, moreover, to obtain proper care for depression than for an ulcer. The other side of this has to do with a recent phenomenon striking at the heart of Canadian health care: doctors refusing to treat physically ill patients due to their vices, such as nicotine addiction or overeating. Because of a scarcity of health care dollars, some doctors reason that they must choose who gets what procedure, not unlike the way a transplant team might choose one potential recipient over another because the potential recipients have different lifestyles. You have thirty days to quit smoking or you must find another doctor. The slippery-slope argument leading to an effective apartheid is not far away, not to mention the foul stench of eugenics.

Returning to the distinction between physical and psychological illnesses, it is hard to imagine a psychiatrist refusing to treat a patient because the patient fails to break the destructive repetition of their behavior. It is precisely the repetition that is the symptom under treatment. Again, the larger point here is that illness is not some discrete condition, where one is either sick or not sick. And this is because illness is not simply a physical or even psychological event; rather, it is also a thoroughly social, political, and cultural event. Illnesses are not just effects of this larger complex of modernity, but they are produced by it. To understand an individual illness as a symptom of something larger directly engages the most crucial ideological assumption of our current moment: that things, events, and people have an autonomous identity, a root cause that is waiting to be neutralized like so many excised tumors or, for that matter, deleted like so much viral computer code.

At this point we can see the importance of our growing love affair with DNA research—the science that promises to identify exact meaning, solve crimes, save lives, and clarify all metaphysical uncertainty. The mapping of the human genome is doubtless one of the most stunning developments in history, which does not mean that it is not also loaded with a tremendous amount of hype and gratuitous chatter. When James Watson, who was instrumental in discovering DNA and played a key role in the Human Genome Project, was asked whether he was playing god, he responded: "Somebody has to." The reference to god is not simply a glib indication of Watson's atheism, but evinces a more significant worldview.

Relying on the wonders of DNA comes at a price. A perfect example of this is Hwang Woo-suk, the South Korean researcher who went from national hero to globally recognized fraud in a matter of months. Until November 2005, Hwang was considered one of the pioneering experts in the field of stem cell research, best known for two articles published in *Science* magazine in 2004 and 2005, in which he reported to have succeeded in creating human embryonic stem cells by cloning. This success offered the hope of converting a patient's own cells into new tissues to treat various diseases. This is the path to regenerative medicine, when the primitive stem cells are instructed to grow into a healthy lung or unparalyzed legs. Hwang was given millions in federal grants by his government and the responsibility of turning South Korea into a world leader in biotechnology. A national postage stamp circulated celebrating Hwang, and the national airline, Korean Air, underwrote his first-class travel. Things fell apart when a group of skeptical younger scientists began to question Hwang's research. This was followed by the recognition that the images published in *Science* to support Hwang's claims were "Forrest Gumped"—digitally enhanced so as to appear as if the stem cells were derived from cloned human embryos. In fact, they were not. The fraudulence was not about the science; many agree that regenerative medicine is theoretically sound and will become the dominant paradigm in the decades to come. Rather, the fraudulence was more about how science cannot be delinked from geopolitics and culture, not to mention from our fantasies about technology.[33]

Love of Convenience

The global abyss opens around the nuclear family as the normative form, a form that receives its legitimacy from the nation at the very moment when other kinship practices and social relationships are emerging (stoked by globalization processes) and challenging the traditional family's dominance.

Although my application for permanent residency was deemed medically inadmissible, if I were to apply as the spouse of a permanent resident or Canadian citizen then I would be granted status without delay. In other words, medical inadmissibility only applies to economic class applicants, not when applying as a married or common-law partner. But this is only a relatively recent stipulation. In the summer of 2002, the Canadian government found itself in a rather high-profile and costly legal battle with a Canadian citizen whose German spouse had multiple sclerosis and was deemed medically inadmissible. Angela Chesters and her husband presented a constitutional challenge to the medical inadmissibility clause of the existing Immigration Act. The Chesterses argued that the immigration ministry's decision violated the fundamental justice sections of the Charter of Rights and Freedoms—basically that the spouse of a Canadian citizen was being discriminated against due to her disability. The Chesterses eventually lost their case. Ruling against them, the federal court argued the following: "This case is not about disability, but the medical assessment of potential immigrants to Canada within the context of Canadian immigration law. By its nature, legislation governing immigration must be selective."[34]

Immediately following this case, a new immigration act was passed, stipulating that medical inadmissibility would not apply to family class applicants. This crucial shift in policy was influenced by the recognition that the government was losing too much money in legal fees (and too much credibility in terms of its compassionate and humanitarian claims) compared with what the health care costs would be to fund these new residents. There are two issues at stake here: one related to the normative definitions of family, and the other to why family class applications should be treated differently than economic class applications.

Following the South Asian tsunami in 2004, family reunification applications for those in the disaster-stricken region were fast-tracked and given priority over nondisaster applications. In a show of compassion, Canadian officials also waived the tsunami-affected applicants' processing and landing fees.[35] Family reunification has recently taken on more of a priority than economic class applicants, with the number of family class applicants rising steadily over the past two decades and finally catching up with the number of economic class applicants. Much of these changes have to do with the popular sentiment around immigration. In the 1990s, outrage was high over the sensationalized view that wealthy immigrants were buying their immigration status. Taking on a racist hue because much of this outrage was directed against Chinese emigrants from Hong Kong, the upshot was that family reunification was recognized as much more in line with the humanitarian and compassionate identity claims of the Canadian people.

One of the main arguments as to why family reunification must be prioritized is that immigration is recognized as an arduous task; the lack of family support makes for a much more overwhelming immigration experience (and, therefore, requires more social and medical services for the new immigrant and, of course, less productivity for the state in return). It is true that we can also trace this emphasis on the family class back, at least, to the era of the Second World War, when some who fought overseas brought back non-Canadian spouses. No one was about to tell those soldiers they could not return to their home country with their new loves. As Canadian immigration demographics changed in the 1960s and 1970s, with a marked shift from Western Europe to Eastern Europe and Southeast Asia, a move away from family class status occurred. Whether or not this had to do with an undercurrent of ethnic discrimination or the economic needs of the postwar Canadian economy is hard to determine. More significant, however, are the reasons the categories of employment and family are so rigidly defined and separated. Why are employment relationships, enduring friendships, and community-based support groups not considered just as fundamental to human dependency, if not human reproduction? This gets us to the privileged role of the family, how the state functions to ensure such privilege, how such privilege functions to ensure the state, and, finally, how globalization processes both undercut and reinforce all of this. The recent debate over gay marriage helps to tease out the connections.

With so many looking to the state to legitimate gay marriage, it is no surprise that many who are dedicated to gay and lesbian rights feel so uncomfortable. To turn to the state for recognition and to marriage as the form that confers such recognition (both culturally and juridically) is potentially disastrous for those who envision and practice other forms of kinship and sexual relations, not to mention for those who wish to question the assumption that it is the state that should furnish these very norms. Judith Butler puts it this way, "Does the turn to the state signal the end of a radical sexual culture? Does such a prospect become eclipsed as we become increasingly preoccupied with landing the state's desire?" Butler is less interested in answering these questions or arguing for the correct stand on gay marriage than in understanding how the state's role in determining who will be included in the norm (who and what is legitimate and illegitimate) effectively forecloses what she calls the unthinkable, a position "not figured in light of its ultimate convertibility into legitimacy."[36]

For instance, in the gay marriage debate the frame has been defined as the legitimate heterosexual family against the illegitimate homosexual family. But this binary does not even come close to exhausting the field of intimate relations. These other relations, however, are not simply illegitimate, but unthinkable when so much attention is granted to a strict configuration of inside and outside. Defining the negative category in relation to its own positivity is one of the most fundamental ways the center reproduces itself. The state has a stake in what is defined as illegitimate marriage, kinship, and other intimate relationships, and when the illegitimate is defined as gay marriage with the possibility of convertibility into the legitimate, then regardless of the outcome the state's work has been done.

Many argue in favor of legalizing gay marriage as a way of securing social services, economic benefits, and, of course, health care. Forget the symbolic value, many reason, we want the entitlements. But why should such entitlements be based on marriage or other legal contracts, contracts that only the state can confer? Not too long ago, the Republicans in the United States fumed at the so-called family tax (paid by those who are taxed extra because they have children) and the "death tax" (paid by those who are taxed extra when trying to pass down wealth to family members). Those who do not have children or who do not have an inheritance to bequeath are never recognized as being taxed extra for not having children or wealth. Yet, fighting these battles on such a political terrain invariably leads to a cartoon over what is morally right and what is strategically wise.

Things got really messy following the decision in June 2004 by the Ontario court to allow for same-sex marriages. When planeloads of U.S. citizens arrived in Toronto to legally marry (Ontario does not require residency when issuing a marriage license), the question then turned to whether such unions would be recognized back in the United States? Canadian and other foreign marriages are generally recognized in the United States via the legal concept of "comity," which is the informal and voluntary recognition by courts in one jurisdiction of the laws and judicial decisions in another. For example, driver's licenses are recognized in different jurisdictions. In almost all cases a marriage that is legal in one country is respected in another, so that people can be secure in knowing that their family is intact as they travel. When Bill Clinton signed the infamous Defense of Marriage Act in 1996, which explicitly frees states from the duty to recognize same-sex partnerships entered into in other states or countries, he paved the way for the current fight over the global recognition of gay marriage.[37]

The conservative backlash to gay marriage might very well be acted out as a struggle over moral values, but the real significance lies in the power over—and sovereignty of—the nation-state. Might this not be one way to explain the brilliant double standard enacted by neoliberal politicians: at once privatizing public institutions *and* taking on a greater role in determining what is morally legitimate in the private lives of its citizens. Perhaps this is no less contradictory than arguing for a radical reduction of government regulation and bureaucracy while considerably expanding one of the largest government-run institutions of them all, namely, the military. If the second Bush administration, for example, had been true to its political-economic positions, it would have openly supported the privatization of marriage, the opening up of the marriage market to a wider demographic of consumers, the freedom to consume equals freedom as such, so the story goes (at least when it is told to the citizens of nondemocratic regimes). Or we can come at this from the other direction: if the Bush administration had been true to its cultural-ideological positions, it would have openly supported gay marriage because any desire for the nuclear family today is at once a resistance to other forms of collectivization—the real threat to capitalist state power. This is not to argue that the nuclear family is inherently conservative, but that commodity capitalism and state power could not thrive without it, or at least the fantasy of it.

One of the more significant threats to state power today is that global capitalism is not as dependent upon the family (in either its really existing or fantasy forms) as was the modern state. This shift (along with the fact that capitalism has not quite dealt with the shift ideologically) has been crucial in opening the global abyss. The story used to go that with each nuclear family comes an expansion of consumer needs and goods that cannot be shared—appliances, electronics, mortgages. At the same time, the family was a key disciplining institution, the place where prohibitions were reproduced. Many feared (and some hoped) that as the role of the family weakened, economic and ideological forces would follow suit, preparing the ground for substantial, even revolutionary change. Notwithstanding the declining role of the family today, however, consumerism and submission to established power is as high as ever—however much this is stoked by a hybrid of nationalist and global ideologies. The struggle over the traditional family, therefore, is also a struggle over the role of the state at the precise moment when its powers are called into question by nonstate powers.

This point returns us to the importance of the family class in terms of immigration policy. And one place where all of this comes together is in the debate over international adoption. One extraordinary case has to do with Romanian-born Alexandra Austin, who was adopted by an Ontario couple in 1991. Although residing in Canada, the parents (a U.S. cardiologist and an Italian homemaker) were not permanent residents. After living in her new Canadian home for only five months, Austin was driven to the airport, loaded on a Swiss Air flight, and sent back to Romania all by herself. It was at this point she entered the global abyss. Not only did she not possess Canadian permanent residency status or citizenship (because her adopted parents had never applied), but she no longer possessed Romanian citizenship (because Romanian policy is to transfer the status of adoptees to the home of their new parents). Romania considered Austin Canadian, while

Canada considered her Romanian. And in the space between (something on the order of a geographical meantime, or meanspace), this poor girl was not entitled to education, health care, or the use of her original name.

Austin, who is now in her thirties and still living in limbo, is presently suing the Canadian government and asking that she be granted citizenship. She is also asking that the present law be changed so that future adoptees are automatically made citizens. Regardless of whether she was granted U.S. or Italian citizenship based on her parents' status, Canada should have looked beyond the family to other forms of human dependency. But, since a nine-year-old girl certainly does not come under the category of economic class, and since she could not be considered a refugee, there was no legal way to process Austin's status. This case is so brutal it almost seems unthinkable that such a conjuncture could emerge. But this is precisely the point: all new spaces and subject positions are emerging that challenge not only state law and policies, but our own powers of imagination.

One of the great narratives in popular culture is the fraud that turns into the true, for example, the marriage of convenience that turns into real love. But already there is a confusion of levels in this circuit, since the marriage of convenience is a political category while love is an emotional category. There is always a moment in the Hollywood film, however, when the imposters actually fall in love—when the emotional and political come together as one. It is for this reason the single most important criterion for immigration officials to determine the legitimacy of a marriage is time: when the couple fell in love. At this point, evidence (photographs and joint bank accounts) is vital, confirming the extent to which the partners know and care about each other. Arranged marriages, as well as other standards of love (such as the love that might be required to enter into a marriage of convenience), only confuse what is already a thoroughly arbitrary event. But it is precisely this conflation of the emotional and the political (or the moral and the economic) that conveys the current crisis in capitalism as it shifts to a newer stage and logic. This shift, moreover, is accompanied by the temporal shift I call the new chronic.

The time that defines a government-sanctioned marriage is, in fact, chronic time—in which the present only matters when defined by the past and committed toward a long-term future. But the past and the future in this case are only understood within the same discursive logic as the present. Expediency (the so-called crime that frauds commit and for which immigration officials are on the look-out) means getting married for the benefits it grants in the present. This is also why the golden ticket to legitimacy (even more than photographs of past anniversaries) is for the partners to have a child. For the state, there is no such thing as an expedient child.

Lee Edelman argues that what he calls "reproductive futurity" (the dominant view in which our beliefs and desires are geared toward building a better life for our children) functions to depoliticize a present in the name of a future promise—a promise that is the child.[38] At stake here, however, is less some misanthropic or decadent disregard for children than the dominant discourse of temporality that reproductive futurity enables and reproduces. This would not be so bad if the future were indeed politicized. But the sacrifices of the present become endless as the rewards of the future become endlessly deferred. And, therefore, the state employs the dominant discourse of children as an excuse to repress and demand sacrifices of its population.

In Butler's conception, one struggles against the state for a wider (transformed) definition of marriage and ultimately for a wider definition of the human, while Edelman argues that the radical position disidentifies with the state (turning to Lacan's theorizing of the Freudian death drive, on which I will elaborate when producing the concept of the already dead in part three), since the queer is, by its very structural logic, outside any available definition of the human. And this negative position and the homophobic fears that sustain it, Edelman believes, should be tactically embraced.[39] To be radical, in Butler's view, means struggling to expand what is possible in the future, while for Edelman it means struggling to occupy the impossible in the present and forgetting about the future.

What is underdeveloped in both of these compelling positions, however, is the very nature of the future itself. The future cannot be abstracted out of capitalism. In other words, there is a capitalist future that is built right into the present and any noncapitalist future can only be imagined within capitalism itself. Radical difference is impossible to represent, given the structuring limits of the present capitalist situation. This historical trap, however, does not mean that a noncapitalist future is impossible, only that it is unimaginable from the current situation. Whereas Butler focuses on family and future in relation to the present capitalist situation (but in a way that does not submit to the structural necessities of this situation's logic),

Edelman can only think about family and future without any relation to capitalism (and in a way that does not recognize any possibility of radical change to this logic). Imagining noncapitalist forms of family and future without dismissing the necessities of capitalism is an almost impossible task. But it is precisely the intersection of these two desires that is flashed while freefalling in the global abyss.

This Is Not a Representation

The global abyss also opens between the present limits of representation and the future possibilities of what can come but cannot be represented. The abyss, therefore, opens with the shifting systems of representation, both political and aesthetic. The example of how global CEOs are called upon (in place of elected representatives) to represent the interests of citizens (and how certain national leaders hold power over those living outside their nation-states) is just one well-known symptom of this political shift. In fact, what we are witnessing all over the world is the limitation of representational democracy. Just as with gay marriage (in which the intense debate over marriage serves to neglect the fact that marriage itself is transforming, and how such a transformation relates to the transformations of the nation-state), there is a similar disavowal occurring on the order of democracy. It is not a coincidence that all the emphasis on the democratization of Iraq and Afghanistan is occurring precisely at the moment when so many are calling into question the stability of representational democracy itself.

Representation is a key category that generally refers to how one thing (or person) mediates the relation between two others. A poem mediates the relation between language and the world. Money mediates the relation between labor and a commodity. A politician mediates the relation between citizens and the state. An advertisement mediates the relation between the consumer and the company. Representation becomes a dominant category in modern life, especially as capitalism becomes the larger system in which daily life is lived. And because of the shifting dynamics of global capitalism, the limits of representation are erupting in the various realms mentioned above, while at the same time new forms of representation (and new forms of engaging the very problem of representation) are emerging.

Representational democracy emerged with the need to rule over expanded populations and territories. The "rule of all by all" had to find a proper form of mediation that recognized the leap in scale from ancient city-states to larger nation-states. Be they communist or parliamentary leaders, chosen representatives were to ensure that both the welfare of the state and the welfare of each individual were properly represented. Different theories of what kind of elections should be held, and how often, were influenced by whether the state or the individual was privileged. Emphasizing all the people (that is, the state) over the individual citizen would usually result in more distance between representatives and the represented—ultimately leading to the rule of the few, as occurred in so many aristocracies and monarchies. Likewise, emphasizing the representation of individuals (to the extent that people could represent themselves) risked placing the interests of those with more power over the means of representation, trumping the general interest of the collective. This has always been the tightrope modern democracies have had to walk.

What happens, however, when the decisions and laws that most significantly govern one's life are produced in a place where one cannot vote? One of the reasons family class immigration applicants are emphasized over economic class applicants in Canada is that the families who sponsor the applicants are Canadian voters, voters who live in political ridings with specific representatives (whereas the economic class applicants have no political representation). The point is becoming clear that representation (by voting, in this case) is an insufficient way to deal with human needs and rights in a changing (globalizing) political landscape, and thus our mode of political participation has to move away from the practice of "being represented."

On the larger global scale, so much of the world's population is not represented in the decision-making bodies of the world, namely, the U.S. Congress, not to mention the United Nations, transnational corporations, the IMF, and World Bank. During the reelection campaign of George W. Bush in the fall of 2003, two new websites emerged that drew attention to the situation. The Netherlands-based theworldvotes.org and the Canada-based Voices Without Votes argued that non-Americans who are directly affected by U.S. policy deserve to have their voices heard. By voting in an Internet poll, participants advocated a form of direct global democracy, one that might effectively grant one-person-one-vote privileges to everyone in the world. What a dream. But this is a dream that seemed to come true with the election of Barack Obama. This is not to say that votes by non-U.S. citizens counted, but that global public opinion certainly shifted the coordinates of the election. At the same time, this global support for Obama was what almost gave John McCain the victory. This fact requires us to distinguish the crucial difference between a vote and an opinion. What we have at present is a substitute for global democracy in the name of public opinion. For instance, in 2005, 30 million Chinese citizens, angry over Japanese amnesia regarding war crimes, signed an Internet petition in response to the possibility of Japan becoming a permanent member of the U.N. Security Council. This example and others, in which new forms of representation come dangerously close to a numbers game unrelated to equality, can also be read as utopian symptoms of a global representational system that cannot yet emerge.

My contention, however, is not to give up on democracy as such, but to pull it down from its pedestal and replace it with other categories, namely, freedom and equality (categories that are admittedly of a different order, but that could serve a much more democratic function). Instead of point-ing toward democracy as the justification for all action (our bombs are democratic no matter how many innocent civilians they kill, whereas his bombs are undemocratic no matter how unreal they are), instead of repre-senting equality in terms of democracy, we might want to invert the equa-tion and ask not whether policies or governments are democratic but whether they produce democracy. How many people are starving? How many people have access to health care and education? How many people are incarcerated? How many people are free to even question democracy—not in some garden-variety form in which every rock thrown at democ-racy's door is understood as a testament to democracy's inherent health, but a freedom that pushes even further, so far in fact that we open up to the "indetermination and indecidability in the very concept of democracy, in the interpretation of the democratic."[40] The simple argument here is that one should focus on freedom and equality, rather than on what they repre-sent. And it is precisely this argument that returns us to the problems of representation.

Within the cultural and philosophical sphere, this exact argument is rehearsed when arguing for the decentering of representational meaning. Take the well-known case of René Magritte's painting *Ceci n'est pas une pipe*

(1923). In this work, a detailed figure of a pipe exists above the handwritten text, "Ceci n'est pas une pipe." Of course this is not a pipe, but an image of a pipe. The point is obvious, but the force of Magritte's intervention (and Michel Foucault's famous essay about Magritte's work) is that when looking first to what the painting represents in the world forecloses what the painting does—work that has everything to do with calling into question representational meaning itself. This is not to privilege the real pipe over the artistic simulation, but to recognize that the artistic work (Magritte's painting) is real itself. It represents not the object world of pipes, but the immanent world of representation whose effects are just as real, just as addictive, just as pleasurable and dangerous as smoking the pipe itself.

A colleague of mine who headed a medical imaging laboratory in Toronto's main cancer hospital reminded me of Magritte's pipe. He showed me a series of slides he uses when teaching courses on advanced medical imaging. The first slide is of Magritte's pipe and the negating text. He then moves to a second slide with the text, "Nor is this a brain." The third slide is a magnetic resonance image of a brain, followed by a final slide including the same text underneath the brain, "Nor is this a brain." My colleague stresses that this lesson is particularly important for medical imagers (and for the radiologists who interpret the images), because when describing images mathematically, where one uses one set of equations to describe the object and a completely different set of equations to describe the image, it is easy to forget what is what. There is a double materiality going on here: one referring to the image of the brain and the other to the brain—both reducible to mathematical formulas and yet both fundamentally different. In other words, the point is not to remind the students that they are dealing with real human beings, and not simply images, but to remind them that the images do not represent the human; they are real themselves.

My illness was caused by a chromosomal mutation. Chronic myelogenous leukemia results from an acquired (not inherited) injury to the DNA of a stem cell in the marrow. This injury to the stem cell's DNA confers a growth and survival advantage on the malignant stem cell, resulting in the uncontrolled growth of white cells, which leads, if unchecked, to a massive increase in their concentration in the blood. In 1960, two University of Pennsylvania physicians noticed that a chromosome in CML patients was shorter than the same chromosome in the cells of normal patients. Pieces of the chromosomes, which are broken off in the blood cells of patients with CML, switch with each other. The detached portion of chromosome 9 sticks to the broken end of chromosome 22, and the detached portion of chromosome 22 sticks to the broken end of chromosome 9. This abnormal exchange of parts of chromosomes is called a translocation, and the presence of the abnormality is called the Philadelphia chromosome.

Gleevec is believed to work by interfering with the abnormality and blocking it from telling the body to keep making more and more abnormal white blood cells. The medication molecularly targets the mutation, thus sparing normal cells. Prior therapies killed off normal cells. Instrumental to the management of the disease are representational technologies that image the tumor load of Philadelphia chromosomes. Since I was diagnosed, the benchmark and so-called gold standard of what indicates remission has shifted. At first there were three types of remission: hematological, cytogenetic, and molecular—the first involved a test to see if the white blood cells were in check, while the latter two searched for the Philadelphia chromosome. Some people could be in hematological remission while reaching only a minor cytogenetic remission, whereas others might be in a complete cytogenetic remission while the disease remains detectable on the molecular level. What complicated matters even more was that different clinics used different representational devices that produced different results—making it practically impossible to compare results with any accuracy. When I was first diagnosed, a certain machine in Germany took on almost mythical status within the CML community, as it was said to conduct the most sensitive analysis in the world. Some patients would fly there hoping to achieve molecular remission on that machine. With the chance of false positives and new diagnostic discoveries, however, the whole system of representation fell apart. What emerged instead was a log system in which a percentage of detection to an overall tumor load would serve as a benchmark. My point in reviewing this protocol is not only to stress how capricious the representation of CML is, but, more important, to stress how such capriciousness has no place within the narrow confines of immigration law as presently configured.

According to certain representational technologies, my leukemia is totally undetectable. Should this be called a cure? No hematologist is willing to go that far. But this is the only place an immigration official is willing to go, despite recent proclamations that cancer will be "maintained" by 2030 and turned into a chronic (not terminal) illness—and this not from some

A full-page
advertisement
for Gleevec
that appeared
in the *New
Yorker*.

Nixonian politician with a superhero complex, but by the president of the American Cancer Association. At the same time, if I were to survive a bone marrow transplant, I would be considered "cured," even though there is an almost 40 percent mortality rate and an almost 50 percent relapse rate after three years.[41] Another key question here involves the timing of the illness, especially since I was diagnosed after living and working in Canada for over two years. Does this mean I got the illness in Canada? And if I did, was it perhaps due to exposure to benzene or another known environmental trigger of CML? Would this change Canada's responsibility for me? This, of course, is an impossible question to answer—both in terms of public policy and biology. How can the law possibly remain stable around so much instability?

The *New Yorker* magazine ran an advertisement for Gleevec over a period of three years. The full-page ad is organized around a picture of the young woman to the left with a paragraph of text narrating her situation to the right. The text begins, "If her cancer had happened a few years ago, Erin might be dead now. But she was lucky. Novartis put her deadly cancer into remission quickly and completely. . . . Novartis is proud to be the innovative force that's bringing new optimism and hope to patients and their families."[42] This paragraph ends with the popular Novartis slogan, "Think what's possible." Above the photograph is a headline that reads, "Stunning Success," followed by two sentences: "Deadly cancer at 23. Complete remission at 24." On the bottom of the page is the Novartis logo.

The desired effect of the advertisement is clear: Novartis saves lives. No matter what you might think about big pharma, about the nefarious economic imperatives shaping this $550 billion industry, about the fact that the industry repeatedly turns its back on the global south as it covets North America, Europe, and Japan, which accounts for 88 percent of its worldwide pharmaceutical sales in 2004, about the way private pharmaceutical corporations take credit for and exploit years of basic research conducted in public institutions, no matter what creeping suspicions you might have about this shady business, without Novartis Erin would be dead. And so would you if you were one of the unlucky few to be diagnosed with CML.

What cannot be missed in this advertisement is that Novartis is not targeting Erin or even a potential CML patient. If there were any doubt about this, one would only need to remember that when the advertising campaign was launched there was no competition for Gleevec (anyone diagnosed with CML would necessarily be directed to Gleevec as a first-line therapy, despite the fact that the ability to pay for such an expensive drug is out of reach for many). Of course, this is not that unusual, since the logic of an advertisement is not only to sell products to its targeted consuming audience, but to fuel desire in the larger market of consumers. Just like with luxury automobiles, high-end jewelry, or other expensive commodities, the larger group of consumers who cannot purchase these goods—but still covet them—is instrumental in stoking the good's value and, more importantly, in reproducing a more general consumerist ethos. In the case of Novartis's *New Yorker* advertisement, the slogan "think what's possible" works to circumscribe the possible by driving home the drug industry's most self-evident and indispensable truth: we (the private sector) represent innovation and the possibility for life-saving escapes from the terror of unpredictable disease.

During the Winter Olympics in 2006, General Electric (GE) launched a major advertising campaign. Four television commercials and four print ads celebrated GE's role in medical imaging. As Novartis did with the slogan "think what's possible," GE asked us to "imagine the future." Both slogans promised a future in which many of the medical threats of the present will have vanished, leaving younger generations innocent to such medical terrors as cancer, neurological disorders, and organ failure. In one print advertisement, a picture of a radiologist's hand points to a strip of data. Underneath the image is the following: "Doctors look at your medical history. Imagine if they could look at your medical future. . . . We're creating revolutionary new ways to predict potential medical conditions before they even occur. We're helping doctors see into the future. Which will make some heath problems a thing of the past."[43]

One of the television spots begins with a close-up of a baby looking directly into the camera. A voiceover explains: "Welcome to the earth. It's a great time to be alive. You'll learn things in kindergarten your parents didn't know in college. You might take your first road trip in a car that runs on water. And you'll experience what at GE we call Early Health. A completely new way of looking at health care that just might make this the best time to be alive. Oh, and by the way, someone is going to walk on Mars. Maybe even someone you know. Welcome to the earth. From GE. Health care reimagined. GE: imagination at work." At the end of the spot, the camera then moves out to reveal the baby on the back of his mother and looking at another baby on the back of his father. The baby who we thought was looking at us (but who was actually looking at the other baby) is white, while his new friend is black. Not only is the world to come one without certain medical terrors, but it is one, perhaps also as a result of GE's technologies, without racial disharmony.

According to the advertisements, GE's technologies mark a qualitatively different moment in history, a future that will establish a different type of human—at least one whose anxieties will be substantially different than his or her parent's. But who will control and run this future? General Electric and Novartis want to hold power over the possible, which is also power over the impossible—how it is and is not imagined and represented. By representing the way the meaning of its behavior is understood, Novartis sets the parameters of the pharmaceutical debate—indeed, it significantly shapes the larger debate about global capitalism. But, like the crisis of political representation, the stability of cultural representation in advertising is also vulnerable. The circuit leading from consumer to company is hijacked by the world, by a critique of representation itself.

How does one move from the Novartis advertisement to global justice and equality? How does one break away from the one-way logic leading from terror to the terror-canceling heroism of Novartis? One must jump the circuit by asking not what the advertisements mean and to what they refer, but how they work. Like Magritte's painting and my colleague's image of the brain, the representation becomes more illuminating when understood for what it is (a material advertisement that is part of a multimillion-dollar marketing campaign). Likewise, we should ask how Gleevec works on a molecular level, so that it can be shared and accessible to all through generics and other forms. We should not ask what Gleevec means, so that it can help preserve and excuse an absurdly profitable industry and inequitable economic system. And this is precisely what is happening as the mediations of everyday life are breaking down and new mediations are being produced. But the new mediations have yet to come, which leaves us in the global abyss between the limits of the present and the possibilities of the future.

The Mediations Have Been Shortened

This returns us to the global abyss and to shifting regimes of power that open up new spaces of both possible and impossible action. Since my diagnosis, and from the very first moment it was recognized that my health condition might present obstacles to my immigration to Canada, I recited the following mantra: I am clean and must remain clean so that if Canada kicks me out of the country it will be solely because of my medical condition. I reminded myself that this narrative must not be sullied by my violating the law or by any other act that can be construed as a transgression. In other words, at various moments in this whole ordeal I had opportunities to settle things once and for all. I could have not disclosed my illness, which would have worked, since the blood test required by Immigration Canada would not have revealed any abnormalities. I could have entered on a spousal application. I could have represented my illness in a way so as to exploit the ambiguity of whether or not it warrants inadmissibility. No doubt everyone would have been relieved (the bureaucrats at the ministry of immigration and the administrators at the University of Toronto) if I had just taken such liberties and simply slipped in under the radar. But this is where things broke down. If I had acted consistently with my own thinking about the logic of crisis and capitalism, I would not have worried as much about defiling the purity of my narrative before the ministry of immigration. Why should I obsess over being clean when the system itself is structured to be so dirty?

This is precisely the point where guilt as a juridical category enters the picture. The unavoidable psychic guilt produced by global commodity culture is, in fact, as inextricably linked to guilt and innocence as the function of these terms under the law. Not only might we try to hide a guilty conscience, but we might also try to hide our guilty acts—though these acts may exist solely on a fantasy level. No matter that one is innocent when searched by a police officer, there is still this lingering fear that something will be found, that some crime will be discovered or planted, that you will be found culpable for an offense that was never committed (and this says nothing about racial profiling or other procedures that presume guilt). And if you are accused of a crime and insist on your innocence and try to expose the error, the more guilty you appear—but guilty not only in the eyes of the officer (the law), but guilty in your own eyes (and this is where the circle closes as the psychic somersaults spin out of control). As we saw regarding property and piracy laws, even this deeply personal guilt must be resisted and a radical hypocrisy mobilized. However, even if we can find the hypocritical courage and analytical clarity to escape this psychic cage, then there is the law to reckon with—a law that is thoroughly contingent and always open to exceptions (with states of emergency, martial law, or "detainee" camps), despite the fact that the law must always seem absolute, objective, reasonable, and neutral.[44]

When applying for permanent residency in Canada, the two tests one must pass are juridical and medical. Police checks must be conducted in all places one has lived for more than six months. Any felony conviction or guilty verdict usually results in an inadmissible judgment. And, as was the case with my immigration application, any questionable medical condition can result in the very same negative judgment. At work here is a dangerous conflation of the judge and the physician, the law and biology—a conflation that in many ways encapsulates a modern form of power at work throughout the twentieth century. Many thinkers have tried to explain the operations of modern power by stressing how power works not only by disciplining us in the usual spaces of authority, but by shaping and controlling the very forms of everyday life, from the way we walk down a city street to the way we laugh at a joke. The private spaces of personal autonomy (our homes, our dreams, our bodies) are where the real work of power is done, so much so that by the time we arrive at our offices or schools no one needs to tell us to do the right thing. This transition from the political to the biopolitical is the defining quality of modern power, in which humans are no longer the objects of political power, but its subjects.

Today there is a new transition under way that might be best characterized as a transition from the biopolitical to the bioeconomic. This is where control over the forms of life that served a primarily political project (producing subjects who actively participated in their repression and depoliticization) passes to a place where control over the forms of life serves a primarily economic project (justifying the conflation of the law and medicine based on an economic logic of scarcity, sustainability, and profit). To be sure, the biopolitical always served a larger economic function (either to sustain the economies of national socialism or those of capitalist nation-states), just as the bioeconomic serves a larger political function (to set in motion an ideological justification for the lack of democracy). Indeed, to invoke the biopolitical is at once to imply the bioeconomic. Still, there remains a subtle but crucial difference, one that can be historically located within the reconfiguration of the geopolitical system and the more recent neoliberal consolidations of the global capitalist system. This difference can produce a fresh perspective from which to view the general transformation of power under way today.

I would have received my immigration status without delay if it were not for the exception of my medical condition. In a different context, we might understand this decision regarding the value and nonvalue of biological life as a sign of dangerous state power. Here is a quote from the specialists responsible for the medical politics of the Third Reich, "The National Socialist revolution wishes to appeal to forces that want to exclude factors of biological degeneration and to maintain the people's hereditary health. It thus aims to fortify the health of the people as a whole and to eliminate influences that harm the biological growth of the nation."[45] The principles of this biopolitics are dictated by eugenics, or the science of a people's genetic heredity. In *Homo Sacer: Sovereign Power and Bare Life*, Giorgio Agamben writes,

> The principle of nativity and the principle of sovereignty, which were separated in the *ancien régime* (where birth marked only the emergence of a *sujet*, a subject), are now irrevocably united in the body of the "sovereign subject" so that the foundation of the new nation-state may be constituted. It is not possible to understand the "national" and bio-political development and vocation of the modern state in the nineteenth and twentieth centuries if one forgets that what lies at its basis is not man as a free and conscious political subject, but, above all, man's bare life, the simple birth that as such is, in the passage from subject to citizen, invested with the principle of sovereignty. The fiction implicit here is that *birth* immediately becomes *nation* such that there can be no interval of separation between the two terms.[46]

This fiction between nativity and nationality entered into crisis following the First World War, as the number of refugees significantly increased and new juridical measures allowed for the "mass denaturalization and denationalization of large portions of their own populations."[47] This new person, no longer a citizen and, therefore, no longer able to appeal to any form of human rights, represents, for Agamben, the secret of modern power; for now nation-states are compelled to deal with bare life not as an enemy to be killed or a national subject to be repressed, but as a life to be taken within a state of exception—a life between life and death. For Agamben, this logic of the camp (a space that is opened when the state of exception becomes the rule) is at the same time the logic of modernity, a political situation we are still living today.

What we need to add to this analysis is the economic imperative appealed to by contemporary nation-states when it comes to the suspension of rights. If what marks the supplementary logic of modern biopolitics is the denationalization of citizens and the seizure of rights (whole populations left in the no-man's-land of the geopolitical system), then what marks the emerging bioeconomics is a justification for inequality based on the logic of the market. In other words, when we follow Agamben and view the camp and other spaces of exception not as anomalies belonging to the past but as "the hidden matrix" in which we are still living and on which our democracies rest, we must be prepared to view the current situation (in which people live or die based on whether they have access to available medications and medical procedures) not as something to be reformed by so many good intentions, but as an equally powerful hidden matrix that necessarily compromises any desire for democracy (at least within a global economy that is not at liberty to suspend the logic of the commodity).

Perhaps even more insidious (and closer to the operational logic of how things work) than the biopolitical, the bioeconomic principle does not foreground the desire to stamp out evil or appeal to ethics as an excuse for violence, rather it simply states that it cannot afford to do the right thing. The ideologies work in different ways. In a backhanded way, I view the cold honesty of this bioeconomic recognition as precisely what is most encouraging and promising today. Likewise, what is most discouraging and perilous (even cynical) is the wishful thinking that the way things work is not so bleak. Immediately following 9/11, there was a glimmer of hope that such recognition might emerge. The hard core of modernity, the chickens coming home to roost, the logic of inequality and terrifying violence, was recognized as part of the system—now they (we) know how the rest of the world feels. But this was immediately recuperated by the redeployment of liberal ideology as the only way to stamp out evil and spread democracy in the name of the free market.[48] But with each global catastrophe and crisis there is a steady weakening of this ideology, while at the same time the bioeconomic principle is strengthened.

In part one of this book, I wrote about guilt and conspiracy and the way established power instrumentalizes guilt as a way to placate a population. Freud argued something similar in *Civilization and Its Discontents* (1929), when looking at how established power represses certain aggressive instincts, thus instrumentalizing these instincts in order to produce guilty subjects who are managed by moralizing narratives of loving one's neighbor. But the bioeconomic seems to shift the ideological requirements of advanced capitalist states. The bioeconomic is not necessarily about the absurd profits engineered by global pharmaceutical corporations, nor is it necessarily about the black-market trade of human organs—both of which are significant symptoms of our current moment. Rather, the bioeconomic is an emerging global ideology, an ideology that exceeds various nationalisms and feeds off of the current logic of contemporary capitalism.

With the emergence of biotechnology and the dominance of the global pharmaceutical industry, who lives and who dies comes down to simple matters of affordability and access that cut across national borders. This hard fact no longer requires long-distance political mediations in order to conceal it, let alone to justify it. Rather, the mediations have been shortened and the inequalities deconcealed as a simple economic analysis suffices to justify human suffering: "Sorry, we simply can't afford to save your life," the dying are told. The really sad part is that this excuse is not a lie, but rather the truth—a truth that seems more and more visible today than at any other time in recent history.

What we know about the global pharmaceutical industry, for example, is that it holds various national health care systems hostage to its ridiculously inflated cost structure and its relentless attitude of snuffing out (or buying off) any attempt to produce cheaper (or even noncommodified) drugs outside of the rigged structure that it controls. It is not as if the pharmaceutical corporations could solve the problem simply by acting more generously, by acting less out of a desire for profit growth and more out of the desire to save lives. They are structurally barred from doing so within capitalism, and if they did substantially change their ways, then we would quite simply be in a different economic system. And it is for this reason criticism of the pharmaceutical corporations can only go so far. They, too, are subject to a systemic problem in which the production of drugs within a globalized commodity system necessarily generates access problems and other undemocratic outcomes. The point here is that as the global bio-economic logic is strengthened and the ideological justifications for injustice and violence shift—they get closer to the objective situation, they become more candid and honest.

At the current moment, the regime of power is shifting to one in which politics will be at the mercy of a larger economic logic. But is this not already apparent in the movement from the First Iraq War to the Second? Back in 1991, the oppositional slogan "no blood for oil" was viewed as a bit too conspiratorial by many; by 2003, however, even those directing the war admitted that the economic imperatives were crucial as they were inextricably linked to "our democratic way of life." Asked why a nuclear power such as North Korea was being treated differently than Iraq, where no weapons of mass destruction had been found, Paul Wolfowitz (U.S. deputy defense secretary at the time) said: "Let's look at it simply. The most important difference between North Korea and Iraq is that economically, we just had no choice in Iraq. The country swims on a sea of oil."

The Canadian government argued against my immigration admissibility due to economic concerns. If I did not convince the justice ministry that there was an error of law then the decision could have been overturned following judicial review. My last opportunity for appeal would have been to ask for an exception—a special minister's permit or an exception to the exception of medical inadmissibility. This exception would be based on a humanitarian decision, not a political or an economic decision. This separation between humanitarianism and political economy represents the precise crisis of the globalized nation-state, for humanitarian acts (by individual states or humanitarian organizations or wealthy philanthropists) are only conducted on the level of the exception, not in any way that can significantly challenge the larger political and economic rules of the current world-system. It is at this point, as the problems shift from individuals to groups, that humanitarian organizations reveal themselves as utterly powerless.[49]

Freefalling in the Global Abyss

The conflicted logic of the larger geopolitical system produces the global abyss. But the global abyss is not only inhabited by migrants, refugees, undocumented workers, the stateless, and others seeking asylum. It is home to the foreigner in general. And the foreigner includes not only all those away from home, but everyone—citizens and noncitizens alike—as we hold on for dear life as history performs its usual somersaults. This inclusive understanding of the foreigner is crucial to our remapping global power and the global system today.

In the long history of U.S.-Canadian diplomatic relations, not a single American has received refugee status in Canada. The primary reason for this is that the U.S. system of judicial review is perceived as strong and fair. This, in fact, was the reason provided by the hearing officer who judged against the refugee application of Jeremy Hinzman, a U.S. soldier in the 82nd Airborne Division who deserted his unit and arrived in Toronto in January 2004 seeking refugee status. Hinzman's lawyer cited the Geneva Conventions on war, as well as the Nuremberg Principles that maintain a soldier's right to disobey illegal orders, in particular orders forcing soldiers to commit war crimes. Since many have questioned the legality of the U.S. invasion of Iraq, most notably the United Nations (especially after confirming the fact that Saddam Hussein did not possess weapons of mass destruction), Hinzman argued that his fighting would also be illegal. Before the Immigration and Refugee Board could decide on the case, however, the Canadian government intervened by arguing the legality of the war must have no bearing on the board's decision. Finally, in April 2005 the hearing officer ruled against Hinzman, asserting that he did not fit the definition of a refugee facing persecution for his beliefs.

Interestingly, soldiers from both Iraq and Iran have been granted refugee status in Canada using a similar argument as Hinzman's. One soldier, a Yemeni citizen serving in the Iraqi army, refused to participate in Saddam Hussein's invasion of Kuwait, while the Iranian soldier refused to participate in chemical weapons attacks. Along with these decisions, many point to old cases involving Sitting Bull's Lakota Indians and runaway slaves in the nineteenth century, or to the most important modern precedent set by more than 50,000 draft-age U.S. citizens who either evaded or deserted military service by coming to Canada between 1965–1973. At the time, Prime Minister Pierre Trudeau was credited as saying: "Those who make the conscientious judgment that they must not participate in this war . . . have my complete sympathy, and indeed our political approach has been to give them access to Canada. Canada should be a refuge from militarism."[50]

Draft dodgers and Vietnam War deserters never received refugee status in Canada, only support once having crossed the border in the form of work visas, permanent residency status, and protection against extradition. But that was back in the day when immigrating to Canada for most Americans was as uncomplicated as presenting a job offer to a border official and receiving residency that very day. Today, with more than eighteen months of processing time required, and with the need to apply from a port outside Canada, asylum seekers such as Hinzman are at greater risk.[51]

At work here is a similar contradiction that we saw between the need for global labor flexibility and the rigidity of national immigration policy. On the one hand, Canadian Prime Minister Paul Martin's government refused to go along with the "coalition of the willing" and made a show of Canada's sovereignty by opposing Bush's foreign policy decisions, but it could not call into question any structural contradictions in U.S. institutions—especially its legal or military institutions. There really is no significant difference today between a Vietnam War deserter and a deserter from Iraq. And this is not simply the case of a weak Paul Martin or Stephen Harper versus a strong Pierre Trudeau. Rather, this is more about shifts that have occurred in the nation-state system as it confronts a growing global logic of integration and fragmentation. This is not to suggest that national differences do not mean anything anymore, of course they do, but they are limited to surface disagreements that become sensationalized on the front pages of our national newspapers. While the back pages (not the editorials or opinion pieces, but the hard news of business and culture) tell a much less confrontational story of partnership and acquiescence. Between national law and international law, the global abyss opens wider and wider.

What is to be done within the global abyss? I have tried to make clear that the abyss opens (in different ways than we are used to) with the new conjuncture composed of political, economic, psychological, medical, philosophical, and cultural qualities. It is intensely personal and inevitably impersonal. It is thoroughly politicized and in excess of the current political logic. And one of the most critical categories that cuts across these different territories and aspects is that of death. Just as the foreigner must be understood much more inclusively, so must the dying and the dead. Moreover, to be killed without dying might not be a uniquely contemporary experience, but this condition—what I call the already dead—momentarily flashes the impossible: a radical alternative to making sense of, and acting within, the current situation. If it is true that we must always die alone, that death like birth is a singular human experience that we must go through, then might the already dead be able to transgress this seemingly human limit and, in a sense, die—and in the meantime live—together? Might the already dead make room for us to collectivize *in this world* on the way to making revolutionary historical change?

PART 3

The Already Dead

Following the global economic crisis of 2008, led by the plummeting of the U.S. residential mortgage securities market, investors were searching for the next profitable asset on which to speculate. This time they would want to hedge even more, meaning they would want to offset their exposure to risk, to the simultaneous freefall of interconnected products. In fact, the sophisticated real estate securities that were bought and sold around the world were supposed to minimize risk precisely by bundling together a diverse set of assets. If one asset in the bundle weakened (mortgages in Florida, for example), others were expected to hold steady (such as mortgages in California).[1] Against expectations, however, home prices dropped en masse, transforming the bundles into junk and exposing the complicity of economic and political leaders—from rating agency executives and investment bankers to government officials. The lesson for many investors was that they did not hedge hard enough. The ultimate hedge would have been to invest in the one asset not directly tied to the larger capitalist market and not immediately exposed to broader economic risks—namely, death. Indeed, death has recently been discovered to be the only product that promises a near perfect hedge. The buying and selling of other people's deaths, what is sometimes called "stranger-owned life insurance," has already grown to a $9 billion industry and is poised to become the asset class of choice in the future, with an estimated market worth almost $26 trillion.

Twenty-six trillion dollars is the amount of life insurance policies in force in the United States, with up to $500 billion of these policies especially appropriate for this market.[2] If a life insurance holder decides that he is no longer in need of the insurance policy, or if he is in greater need of an immediate payout before he dies, then he can sell his policy to a firm that will bundle and resell it in the same way Wall Street firms sold subprime securities. In the past the "cash surrender value" that the insurance company might pay out to a client who gave up an existing policy was only a fraction of its value (say, $33,000 on a $8 million policy). According to the new scheme, a life settlement company would buy the insurance policy from the client (for approximately $600,000 on the same $8 million policy) and would then bundle policies together and sell them to investors the world over.[3] These investors would continue to pay insurance premiums on each policy and get paid when one of the original policy owners dies. The sooner the policy owner dies, the more money the investor makes. If the original owner lives well beyond their actuarially determined death date, the investor would make considerably lower returns and have to wait years to collect.

Like firms involved in mortgage securitization, life settlement companies must use complex mathematical formulas to configure a bundle of policies that minimize risk. This is done by diversifying the type of illnesses represented in each bundle. If too many policies are attached to a single disease, such as a certain form of leukemia, that is suddenly curable or manageable in the future, the investment would be a bust. The new management of formerly terminal illnesses threatens this type of investment—especially if the new paradigm of targeted therapies can be successfully mobilized across a variety of illnesses, thus compromising the diversity of the investment package. Still, what makes this investment unique and particularly attractive is that, unlike any other product, death is not directly tied to the capitalist economy. Bank runs, overproduction, underconsumption, privatization, and other defining aspects of the capitalist logic do not necessarily influence the year or even the decade during which a person will die. Different modes of production do shape life expectancy differently (the specific economic and social stresses associated with capitalism, for example, can generate certain diseases that will be different than those generated by feudalism), but usually people do not immediately live or die based on the rise and fall of commodity prices.

The cheerleaders for this new market argue that not only does this unique investment promise untold fortunes, it provides an honorable service to those in terrible need. Critics are afraid that a vulnerable senior population is particularly susceptible to predatory buyers, and that the logic of the investment might undermine health care policy and elder care. If the insurance settlements market grows and becomes mainstream, how might this influence (however indirectly and unconsciously from all involved) euthanasia policy or coma care protocols? And there is the problem of alleged bid rigging. There have already been several lawsuits accusing life settlement companies of paying off competitors with the aim of terminating the escalation of bids.[4]

The question that does not get asked in this emerging debate is why someone would want to sell off his or her life insurance policy in the first place. Why, in other words, might they be so desperate for an immediate return? The most common reason in the United States has to do with exorbitant medical fees and the lack of universal health care, not to mention the failure of the welfare state to provide for the most basic human needs, for example the needs of family members who survive the deceased. Life settlement companies emphasize the other side of this: the ideal policyholder is one who no longer needs her policy because she has outlived the risk against which she was insured in the first place, or a spouse has passed away. Life insurance has existed for centuries, but what is markedly different in this new life settlement market is that the investor is a stranger, someone who has no relationship to the individuals whose policies they have purchased.

This commodification of individual death (of the end of life) has an interesting counterpart in what might be called the commodification of socioeconomic death (of the end of capitalism). Right before the worst of the economic crisis hit in the fall of 2008, Donald MacKenzie wrote an article in the *London Review of Books* titled "End-of-the-World Trade."

MacKenzie observed, "The trade is the purchase of insurance against what would in effect be the failure of the modern capitalist system. It would take a cataclysm—around a third of the leading investment-grade corporations in Europe or half those in North America going bankrupt and defaulting on their debt—for the insurance to be paid out."[5] This "revolution insurance" entails companies buying and selling protection on the safest investments possible. What puzzled MacKenzie is that regardless of how unlikely such a collapse is, the cost of insuring against it had increased ten-fold in only six months before he wrote his article. Why would firms buy such insurance, moreover, if they knew that the very firms who sell this revolution insurance will almost certainly be part of the wreckage, not being in any position to make payouts after capitalism has busted. And if there is a great ecological event or even world communism that marks the historical break, then the global economy will be so radically changed that value itself will be utterly reconfigured, turning every asset upside down. MacKenzie explains, "They are looking to hedge their exposure to movements in the credit market, especially in correlation. Traders need to demonstrate they've done this before they're allowed to book the profits on their deals, so from their viewpoint it's worth buying protection from bond insurers, even if the latter would almost certainly be insolvent well before any payout on the protection was due."[6] A company purchases this type of insurance, therefore, not to protect against an accident or an unexpected future, but to better position itself in the known present, in the period before the potential world-changing event. In this instance, insurance can only protect the insured in the present, for in the transformed future the insured, the insurer, and the very category of insurance itself, will be something altogether different.

The commodification of death (both the death of individual humans and the death of the socioeconomic system) teaches us about two crucial aspects of the relation between life and death. First, death is already a part of life—it is already calculated into the value of life. In the first example of insurance settlements, a person is able to sell her death while she is still alive; while the promise (however improbable) of the second example of revolution insurance is that a firm is able to safeguard against death, so that when death (the revolution) arrives the firm will be able to continue living as if death had never arrived. At the same time, these two examples reveal how death is something ultimately separate from life in a way that precludes any integration of the two conditions. Death insurance (both individual and socioeconomic) reveals how death stands at a future, uncrossable distance from life, suggesting the possibility of a radically alternative value system, but one that cannot be fully imagined or realized in the present.

The Always Dead, the Always Already Dead, and the Already Dead

These two ways of representing the relation between life and death (that of assimilation and separation, identity and difference, continuity and discontinuity, even synchrony and diachrony) that we see at work in the two examples above serve to reintroduce the concept of the already dead. The already dead enable a rich engagement with these two opposing modes of thought, one mode removing the separation dividing life and death and the other retaining their relative autonomy. Both of these modes aspire to the same goal: to reclaim our own deaths as a way to reclaim our right to pursue alternative personal and political desires. Reclaiming our own deaths, not in a suicidal way, but in terms of our emotional and political consciousness regarding death and dying, therefore, is tied to the reclaiming of utopia. And by pursuing the problem of utopia, or the problem of the impossible, we are at the same time pursuing the problem of the already dead, for it is impossible to resolve the antinomy of assimilation and separation, identity and difference, life and death. The already dead suggests this impossibility, but it does so in a way (and now we get to the real stakes of this concept) that is specific to our current historical moment, for the antinomy of life and death that the already dead engages has been reconfigured within the context of late capitalism.

The "already dead" names that state when one has been killed but has yet to die, or when one has died but has yet to be killed. But the already dead (although suggesting the impossible, which, as such, persists throughout history) is a subjectivity that is specific to the contemporary moment. And it is this historical component that will help distinguish the already dead from two other related, but less historically contingent, ideas: the always dead and the always already dead.

Let's begin with the always dead and the most famous invocation of this phrase: god is dead. In his seminar on ethics, Lacan digressed to tell an anecdote about a dream of a friend and patient of his. This friend, agitated by Lacan's seminars and style of working through ideas, cried out, "But why doesn't he [Lacan] tell the truth about truth." "If [I] did tell the truth about truth," Lacan explains, "then not much truth would be left." At this point, Lacan turns to god. "We know that God is dead, this is the first truth, but what is the truth about this truth." Lacan continues, "The next step is that God himself doesn't know that. And one may suppose that he never will know it because he has always been dead. This formula nevertheless leads us to something that we have to resolve here . . . that *jouissance* still remains forbidden as it was before, before we knew that God was dead."[7]

"Jouissance" is Lacan's term to describe the mixture of pleasure and pain we aspire to in our drive to transgress limits, limits that are impossible to transgress since what exists on their other side is not only prohibited but impossible (or prohibited precisely because it is impossible). At the same time, the transgression of these limits, for Lacan, has something in common with how we have been arguing for a renewed relationship to death, a reclaimed right to die. For Lacan, attaining jouissance is impossible because the subject enjoying jouissance is a nonsubject. The subject is only a subject negatively—that is, to the extent that she defines herself against things she does *not* do, limits she does *not* cross. But Lacan argues that any reconfiguration of the subject is possible only when the subject is willing to risk its own undoing, its own death. His great example (in this same seminar on ethics) is Antigone, who is "between life and death" or "between two deaths." She participates in life only, as Lacan says, from the perspective of having already lost it. It is while in this unique position between two deaths (having ceased the effort of self-preservation) that Antigone offers her radical critique of Creon's social order. While standing directly on this limit between life and death, Antigone can upset the chain of signifiers through which erotic energy is channeled.

Returning to the death of god, however, Lacan reminds us that merely knowing that god is dead and that he has always been dead does not change a thing. It does not immediately open up a new relationship to the world. In fact, it might be this very knowledge that god is dead that is the most crucial support for the very existence of god. The point is that truth still matters (and exists), but is most effectively approached (even desired) by way of a movement through the various truth effects—nothing changes by knowing the truth about truth.

Now we might want to ask, What shifts when we place ourselves into the pronoun position of the phrase "We are always dead"? But we ourselves do not know this and we can suppose that we never will know it because we have always been dead. But jouissance is still as forbidden as it was before, before we knew that we were dead. Does the analogy work? It all depends on how we understand god and ourselves and on what level our understanding is taking place. It might not be too perverse to understand that we have always been dead if we understand this "we" as something that refers to a certain symbolic understanding of humans, or even to what is called "the death of the subject." The always already dead introduces another temporal aspect: once we understand that we are dead, we then retroactively realize that we have always already been dead. "Our past begins to change," as Bruno Latour put it when discussing the "always already" in terms of our experience of the modern.[8] But why stop here? With this new understanding of our symbolic death, our future begins to change as well.

The concept of the already dead takes us beyond the symbolic, beyond the death of the subject. Here again, we can follow the Lacanian categories, in his use of what is known as the three orders: the Imaginary, the Symbolic, and the Real. For over fifty years, from his early work on the mirror stage in the 1930s all the way to his famous dissolution of the École Freudienne de Paris in Caracas in 1980, Lacan pursued these three realms and the relations among them. All the while, the realms themselves were changing (not just because Lacan theorized them differently at different moments of his seminars, but because of how the history during which they were being thought was changing). The emergence of the Imaginary, the Symbolic, and the Real, therefore, is inextricably linked to the sociopolitical history that overlays Lacan's work, a history beginning with the moment before the Second World War, the war itself and its aftermath, the great social confrontations of the 1960s, the Cold War, and the beginnings of neoliberalism ushered in by Thatcher and Reagan. The emergence of the realms is linked no less to the intellectual history of Lacan's time, the radical theories of structural linguistics (those of Saussure and Roman Jacobson, in particular) and the whole structuralist revolution pursued by Lévi-Strauss, the existentialism of Heidegger and its Marxist inflection with Sartre, the feminisms of de Beauvoir and Luce Irigaray, the pervasive influence of Althusser and his progeny, and the full-blown poststructuralisms of Foucault, Deleuze, and Derrida. This intellectual history must also include its cultural component: the surrealists, the situationists, not to mention contemporary rereadings through these developments of so much premodern and early modern culture, from Sophocles to Shakespeare to Sade. The transformations within psychoanalysis itself were most influential: Freud and the medicalization of his work in France, Jung, the International Psychoanalytic Association, the ego-psychologists, Klein and the various Lacanians, from which Lacan was only too happy to distance himself. "It is up to you to be Lacanians if you wish; I am Freudian," Lacan contended. To remember these histories is to remember that the Imaginary, the Symbolic, and the Real are not stable realms, as we will see with their mutations when we push them into the present moment and in relation to the present transformations of death.

The Imaginary is that realm defined by the content that makes up our lives, the people, the plots, and the particular objects. We understand ourselves in relation to this content—so that my dissatisfaction is because of this person or that event, and if only I or they or it would behave differently then I would be happier, satisfied, and done with all of the alienation. Although there is a certain truth by way of the other (our own truth reflected back to us in the mirror or in our mother's eyes), the Imaginary is based upon the intrinsic identities and values granted to everyone and everything in our lives. It is for this reason that the strengthening of the ego (as a defense mechanism or as the foundation of the empowered individual) is the gold standard of health and normalcy within the Imaginary. In this realm, death functions like an evil person, dressed in black and coming too soon, or sometimes as a savior, dressed in white and putting us out of our misery. The way cancer is commonly personified as an enemy to be defeated or as a thug who needs to be put in his place (or simply eradicated, deserving the death penalty) exemplifies death in the Imaginary.

We might also be reminded of how this relates to the new dominance of medical imagining. Magnetic resonance imaging, PET scans, and ultrasounds show us the content of our deaths; the tumor is isolated and our impending death is confirmed and quantified. The Imaginary is, therefore, also the realm of continuity, of a linear narrative that coheres and develops over time, like the imaged tumor. Likewise our deaths have an epidemiological explanation that can be tracked to inheritance or contagion or random mutation. We might not know why we were unlucky enough to become deathly ill, but we almost always find a name for our illness—even if what kills us has no explanation and no name, therefore taking on what in such cases goes by the name Syndrome X. And this process of naming relates to another aspect of death in the Imaginary, how once we name something the thing itself is effectively killed, losing its self-rule as it is now stuffed with its new name.[9]

The physiological autonomy of cancer is impossible to separate from the name "cancer." And the medical image of cancer functions not unlike the name. Here we might be reminded of recent criticisms of medical imaging in which the highly sensitive imaging technologies reveal cancers that are then communicated to patients whose health actually deteriorates faster due to the image and the name. And there is the example I gave of my colleague who began his seminar on medical imaging by linking the famous Magritte painting *Ceci n'est pas une pipe* to a slide pairing an MRI-imaged brain with the line "nor is this a brain." My colleague was hoping to remind the students that all of the complex mathematics and physics that make up the practice of contemporary medical imaging still have a referent—real patients with real illnesses. Actually, the image *is* a brain, just as the brain is an image—each stuffed full of the other. Here we might make a play on Lacan's joke about the madman and the king.[10] What is crazier than a king who thinks he is a madman? A king who thinks he is a king. The king has no essential kingness. He is only a king by way of his subjects, who treat him as a king. And any king who does not understand this operation of meaning is mad. And now back to our medical image of the brain and the brain itself. What is more ridiculous than an image that thinks it is a brain? A brain that thinks it is a brain.

The Symbolic is the realm of differential relations, of negativity, whereby the only entity granted positive or intrinsic meaning is the symbolic structure itself. Entities within this realm only have meaning in relation to other entities within the structure, and ultimately in relation to the larger systemic logic. In the Symbolic, content is arbitrary and less significant than the formal relations that exceed every individual person, thing, concept, name, or image. In this realm, death structures the relationships among all the institutions involved (law, family, religion, immigration, economy, culture). It is only when someone actually dies that the constancy is complete, that the contradictions among the institutions (such as the need to restrict immigration entitlements while expanding the influx of immigrant workers in service to the capitalist economy) are resolved.

We can return to the issue of medical imaging in order to emphasize the relation between the Imaginary and Symbolic realms (and to anticipate the third realm of the Real). Medical imaging has become instrumental to contemporary medicine. Ideally, medical images properly expose the truth of the medical situation: the tumor, for example, is accurately represented in the image, correctly interpreted by the doctor, and correctly communicated to the patient as a positive diagnosis. This is what we can call the "positive-positive." Similarly, there is the "positive-false": when the doctor correctly interprets an accurate medical image of a nonexistent tumor, and this results in a correct diagnosis. In this case we exhale a sigh of relief after our doctor explains that there is nothing there.

Sometimes, however, the medical image picks up "too much" or it exposes a tumor that does not exist. This is called a "false-positive": the doctor incorrectly interprets an accurate image of a tumor that actually does not exist in the patient, and this results in an incorrect diagnosis—you are told you have cancer when you actually do not have it. These three diagnoses represent the limits of the Imaginary: we get it right or we get it wrong. These possibilities are illustrated on the following page.

Within this realm there is a frustrating dependence on the image. And like interpersonal relations in the Imaginary in which our self-identity is structured by way of someone else, in this case the external medical image shapes our very health (not only in terms of the psychological consequences that occur after hearing about an illness, but in terms of the actual physiological consequences, when, as mentioned in the example above, actual changes to a tumor might occur due to the knowledge of it). The limit is not simply that there might be a false diagnosis or that we might be terrorized by demands made by the promise of medical imaging itself (obsessively consumed with whether it is too early to get imaged, or too often, or too expensive), but that medicine is becoming utterly dependent upon medical imaging, and thus upon the unstable and arbitrary image itself. Medical imaging, moreover, can only image the progression of an already existing tumor or of an already nonexistent tumor. It cannot image the future emergence of a yet-to-be-detected illness. Or, more significant, it cannot image a future in which illness itself functions differently. And it is with this impossibility that we reach the Symbolic.

Interpretation	Medical Image
Tumor	Diagnosis

Positive-Positive

(Interpretation) Correct	(Medical Image) Accurate
(Tumor) Existent	(Diagnosis) Correct

Positive-False

(Interpretation) Correct	(Medical Image) Accurate
(Tumor) Nonexistent	(Diagnosis) Correct

False-Positive

(Interpretation) Incorrect	(Medical Image) Accurate
(Tumor) Nonexistent	(Diagnosis) Incorrect

An illustration of the relationships between the imaginary and symbolic realms.

What about a "false-false" diagnosis? If we remain in the logic of the former diagnoses, then in this case there is an incorrect interpretation of an inaccurate image of a tumor that actually exists, but this misinterpretation results in a correct diagnosis. There actually is a tumor, the image inaccurately shows no tumor, the doctor misinterprets this to mean there is a tumor and tells you that you do have cancer, which you actually have. The play of negatives in this sentence is almost impossible to get our heads around, which is not surprising since the Symbolic is that realm that offers no direct access to its overarching logic. There is an actual structure that shapes our lives, but it is not present and whenever we locate it in one of its particular instances or elements we incorrectly detect the very structure itself. The symbolic structure exists, but it is always wrongly detected, represented, and diagnosed. It is precisely through this failure that the false-false gets as close as possible to the truth of the Symbolic, not by effectively affirming something (two negatives equal a positive), but by banging up against the ultimate limit of the negative structure itself. The false-false diagnosis performs the limits of medical imaging. It can only show us a future within the terms of its own logic, a logic that is based on the unstable image; and it cannot show us a future content beyond the limits of its own form, which is to argue that medical imaging cannot show us the future.

Here we might recall the simplest and most incisive criticism against structuralism: that it cannot adequately theorize change, which means that it cannot incorporate into its formal logic the inevitable quantum leap into a qualitatively different symbolic structure. Or we can say that the Symbolic cannot deal with its own excesses, with the radical breakdown of formalization itself. And when it seems to do so, when it seems to formalize its own radical undoing, it necessarily does so on its own terms and thus

False-False

(Interpretation) Incorrect	(Medical Image) Inaccurate
(Tumor) Existent	(Diagnosis) Correct

A false-false diagnosis.

inadequately. This is not to follow the naive criticism that the Symbolic is ahistorical; rather, it is acutely historical insofar as it is located at a specific historical moment and its entire logic is understood to be based on the specificity of this history. This is the power of the synchronic—that meaning is produced by way of a horizontal move out to present, simultaneous history, while refusing the vertical move back to the past or even forward to the future. Since meaning is made by way of the differential relationships within a specific, historically shaped structure, an entity in a present structure (be it a linguistic unit, a cultural ritual, a nation-state, or a medical image of a malignant tumor) cannot be facilely compared to the same entity in a past structure, given that the logic of the relationships will be qualitatively different and that there is no positive, persisting quality of any entity over time. It is true that we may identify malignant tumors in human bones discovered on an archaeological dig, but the very meaning and effects of that malignant tumor (how it is configured in relation to the dominant political, cultural, and even biological discourses of its day) will be qualitatively different than the meaning and effects that such a tumor will have at different historical moments.

Nevertheless, the Symbolic inexorably tries to incorporate into its synchronic logic a gesture to this disavowed diachronic component. We see this when returning to the false-false diagnosis. The double negative aspect of this diagnosis constitutes a special temporality, one that reveals the very history of change and how the same category can operate differently, depending on where it is located on the string of time. The double negative in speech, for example, attempts to formally capture the irresolvable problem of the synchronic and the diachronic. The first negative is identical to the second negative, but they are also different, due to the changing context. The double negative performs its own limits, just as the false-false diagnosis performs the limits of medical imaging, most significant, that it cannot affirm a content beyond its own form.

And it is right at this most elaborate point in the Symbolic that we jump to the Real, to the third order that informs both the Imaginary and Symbolic and undoes them. This jump moves us from the false-false diagnosis to the positive-positive. Although this positive-positive diagnosis shares the identical form as it does in the Imaginary (a correct interpretation of an accurate medical image of a tumor that actually exists, and this results in a correct diagnosis), the positive-positive in the Real is something altogether different.

Lacan's use of the Real shifts, or becomes more elaborate, in the course of his seminars—beginning with the "real" in the early seminars (uncapitalized and somewhat deemphasized in relation to the Imaginary and the Symbolic), emerging as a fundamental concept by seminar 11, and moving to the forefront of his thinking in what is known as the late Lacan.[11] The Real is commonly understood in three seemingly contradictory ways. First, as that "which exceeds what exists"; second, as that "which resists symbolization"; and third, as that "which is always in its place." Excess, resistance, perfection: these three logics cannot be thought together. One of Žižek's most brilliant moves is to resolve this type of paradox by rigorously delineating the levels, so that any Lacanian term is always triple. For example, there is the Imaginary Real, the Symbolic Real, and the Real Real.[12] As for the three logics of excess, resistance, and perfection, we can now see that the content of the Imaginary is always shaped by its excess—not a repressed content, but a repressed form that ascribes positive meaning to the content (we can associate this with a positivist critique). The Symbolic is always shaped by what resists it—not by more content or another formal aspect, but by the very negation of the form-content binary (associated with a negative critique). The Real is always shaped by its perfection—which also means that it is impossible to hold or represent or formalize (associated with utopian criticism). The Real Real is outside discourse, outside the very category of the outside, however much it necessarily shapes everything (making it, at the same time and in some remarkably impossible way, the ultimate insider). The Real is not what is most real

Positive-Positive

(Interpretation) Correct	(Medical Image) No Image
(Tumor) ?	(Diagnosis) Correct

The positive-positive scenario wherein the category of a tumor transforms.

or what comes after we successfully pass through the other realms, rather it erupts from inside to change everything. Likewise the positive-positive is not what comes after all of the other diagnoses; rather, it erupts to change the very coordinates of the diagnostic process by changing the very coordinates of life and death.

In this positive-positive of the Real there is no image of an unknown tumor. In other words, positive-positive indicates a future in which the meanings and the effects of a tumor are radically different than what they are now and, therefore, they are unrepresentable in the present. It is no longer the case that either there is a tumor or there is not a tumor, or that with the interpretation of some detected DNA information in the present the emergence of a future tumor can be confirmed, but that the very category of a tumor transforms (what it means medically, socially and culturally). Despite there being no image to interpret this unknown and unrealized tumor, there still can be a correct interpretation of this situation and even a correct diagnosis. In this case, interpretation comes close to a kind of belief and the diagnosis comes close to an act of faith—a belief that there is a radical limit to the medical image and that our desperate desire for it to save us from an impending health crisis is repressive. Our diagnosis is correct when we delink our desperate desires from the image and allow our unconscious desire to act for the impossible.

This positive-positive reveals a future beyond the capacity of the imaging technologies to represent. It is an affirmation of unconscious desire, an affirmation of a radically different future. The positive-positive suggests a future that can come, but—at the same time—it is not the kind we can know about, the kind that will accommodate our current ideologies, or the kind that will allow us to behave in our accustomed ways. Images of

(Unconscious) Desire	No Image
(Future) ?	Political

The already dead flashes an unknown future without an image.

illnesses do not and cannot show us truth, because we overwhelm them with our desire for the wrong truths. When we delink these desperate desires from the images and delink these desperate images from our desires, something shifts, enabling images to reveal a future that, although we cannot read, imagine, or directly desire it, is already with us and already more open than anything we presently know. It is when we reclaim our unconscious desire for this unknown future without an image when we have entered, what can only then be called, a truly political dimension.

The eruption of the positive-positive is related to the already dead. In the Imaginary the already dead is the positive-positive as *correctly diagnosed* within medical imaging. Our tumor, clot, or mutation is detected and our eventual end has been projected back onto the present. In the Symbolic, the already dead is the false-false as *incorrectly misdiagnosed*. Our tumor is missed *and* the diagnosis is wrong—which means that we are effectively told that we do have the tumor that, in fact, exists. In this case, the truth of the tumor comes by way of its negative, by the flash that death is something that has come even though it has not been detected. In the Real, the already dead is the positive-positive as the impossible diagnosis, or the diagnosis of the impossible: it reveals a future beyond the capacity of the imaging technologies to record; it affirms a content beyond its own form.[13]

The already dead can be understood as a state that anticipates a different ideological constellation in which death functions. And this new constellation radically reconfigures the relation between life and death, so that the antinomy that I wrote of earlier (death as continuous with life and death as discontinuous with life) shifts as well. This relation between life and death is a profound philosophical and political problem and, therefore, has claimed the attention of some of our most important thinkers—all of whom have engaged the antinomy in one form or another.

First, let's turn to the death drive as mobilized by Lacan and contemporary Lacanians. Lacan notoriously returns to Freud's *Beyond the Pleasure Principle* (written in 1920) in order to claim that the "beyond" in Freud's text is precisely beyond the separation of the life drive and the death drive, so that the death drive is not a separate drive but the pure form of every drive.[14] For Lacan, 1920 marks one of the key moments in psychoanalysis because it is when Freud reappropriates Freud, meaning it is when Freud argues against the emerging positivist chokehold on his thought, one that would later take the name of ego-psychology. This anti-Freudian positivism refuses the importance of the unconscious, refuses the dynamic relation among the various psychoanalytic categories (which is why this is also the moment when Freud develops the so-called second topography—in which Id, Ego, and Superego are constituted within a flat space of interdependence), and refuses history itself, insofar as the modern subject and psychoanalysis can only be theorized in relation to the historical structure in which they exist. By questioning why we repeat behavior that is harmful and unpleasurable, Freud rejects an equilibrium model in which the strengthening of the ego is called upon to manage the different instincts. The death drive, rather, cannot be *managed* without repression and the unconscious repetition of symptoms that allow such management to succeed. To manage the drive is to betray the revolutionary dimension of the psychoanalytic project.

Freud writes in the last paragraph of *Beyond the Pleasure Principle*, "The pleasure principle seems actually to serve the death instincts," and he finishes the essay with the following recognition of his own limits: "This in turn raises a host of other questions to which we can at present find no answer. We must be patient and await fresh methods and occasions of research. We must be ready, too, to abandon a path that we have followed for a time, if it seems to be leading to no good end. Only believers, who demand that science shall be a substitute for the catechism they have given up, will blame an investigator for developing or even transforming his views."[15] Lacan abandons the mistaken separation of the two drives, the life drive and the death drive, that invariably turns into a facile vision of light and dark, good and evil. He famously states in seminar 11, "The distinction between the life drive and the death drive is true in as much as it manifests two aspects of the drive."[16] Ten years earlier in seminar 2 (1954), moreover,

Lacan referred to Oedipus as living a life that is dead, "that death which is precisely there under life."[17] And, most notable, there is seminar 7, in which Antigone's act of properly burying her brother Polyneices positions her between two deaths, the symbolic death and the imaginary death. Antigone is symbolically dead, meaning that she is dead within the symbolic order (she cannot be recognized or be given citizen status precisely because she no longer receives her value in relation to the others in Crete), but not yet dead biologically, or dead as understood in the Imaginary.[18] It is Slavoj Žižek, with his long-standing references to the undead, who best captures this Lacanian inextricability of life and death.

The Undead and the Living Dead

Žižek writes, "The paradox of the Freudian 'death drive' is therefore that it is Freud's name for its very opposite, for the way immortality appears within psychoanalysis, for an uncanny *excess* of life, for an 'undead' urge which persists beyond the (biological) cycle of life and death, of generation and corruption. The ultimate lesson of psychoanalysis is that human life is never 'just life': humans are not simply alive, they are possessed by the strange drive to enjoy life in excess, passionately attached to a surplus which sticks out and derails the ordinary run of things."[19] This undead urge is not desire, but drive. Whereas desire is about a perpetual movement toward an impossible and always-missed satisfaction, drive does not miss anything. Drive always hits its target, since it cannot help but hit its target. But this is the case only because drive's target is a drive for an unattainable goal. Drive is not the expression of neurosis, the misguided, counterproductive, and frustrating search for the impossible, but something more transgressively fundamental and fundamentally transgressive. Drive is the politically and psychologically productive search for the impossible. In this way the distinction between desire, neurosis, and drive resembles our tracing of the already dead through the Imaginary, the Symbolic, and the Real. Desire and neurosis, too, are relentless pursuits of impossible objects (frustrated by excess or resistance); however, it is drive that expresses this logic of impossibility in a way that not only frustrates but opens up new possibilities. It is as drive that the utopian shade of human behavior emerges.

It is this privileging of drive over desire that defines Žižek's radical project. He writes, "A drive does not bring satisfaction because its object is a stand-in for the Thing, but because a drive, as it were, turns failure into triumph—in it, the very failure to reach its goal, the repetition of this failure, the endless circulation around the object, generates a satisfaction of its own."[20] This gesture to the different temporality between desire and drive is crucial. Whereas desire is propelled forward in time, always chasing after a missed object, drive encircles its object and does so in a way that does not move forward in time. Likewise, the death drive is not the progression or retrogression from the life drive, rather the death drive is a combination of the life drive and the death drive. It is one. It is atemporal.

And so is the undead. It cannot become more or less undead over time. The undead is the human when there is no separation between life and death, when the life drive and the death drive, life and death itself, are united. To this nonprogressive, nondivisive mode Žižek ascribes a temporality that is something like a rotary movement (a movement that is not progressing toward somewhere). "This rotary movement, in which the linear progress of time is suspended in a repetitive loop, is drive at is most elementary. This, again, is 'humanization' at its zero-level: this self-propelling loop which suspends/disrupts linear temporal enchainment. This shift from desire to drive is crucial if we are fully to grasp the crux of the 'minimal difference': at is most fundamental, the minimal difference is not the unfathomable X which elevates an ordinary object into an object of desire, but, rather, the inner torsion which curves the libidinal space, and thus transforms instinct into drive."[21]

The zombie in horror films represents the undead within the Imaginary. It cannot be killed. It keeps coming back. It has been killed *and* still lives. It lives *and* is dead. Unlike zombies, however, the undead do not reflect the manner in which they were killed—they do not live forever with axes lodged in their skulls. And this takes the undead from the Imaginary, in which it is invested with so much content (the uglier and slimier the better), to the Symbolic in which the undead is now the place where the structure breaks down. The undead is the structure's repressed—revolution, but revolution not from the outside or that comes after the development and corruption of the structure, but that is already inherent in the structure itself. And, finally, the undead of the Real is the impossible but very real timeless instant that exists between the total breakdown of one structure and the immediate production of another, but that is repressed at the very moment the new structure comes into being.

For Žižek, the undead is not the same as the not dead or the not living. The undead is the living dead that is not external to life and death, but is the excess that is inherent to life and death itself.[22] In other words, there is a difference between life and death. But the difference comes first. This is what Žižek calls in the above quote "minimal difference"—a difference that is prior to the elements that it differentiates. First comes this difference, then comes life and death and, more important, the separation between them. This point is absolutely essential to understanding the Žižekian project (and mapping out how contemporary thinkers agree and take issue with it). The inner torsion that curves space comes prior to the elements that we understand as having curved this space in the first place. As Žižek likes to argue when rereading Freudian trauma, trauma does not curve our psychic space (whereby a traumatic event leads to traumatization); the already curved psychic space traumatizes a trauma (whereby an already traumatized space leads to a traumatic event). The curvature comes before the trauma.

Likewise, the undead comes before life and death. But we invariably think that life and death come before the undead. Let's take the two examples that Žižek cites in order to think about being in between two deaths: Antigone and Hamlet. As mentioned when introducing the Lacanian death drive, Antigone is symbolically dead, but biologically alive. Hamlet's father, on the contrary, is biologically dead but symbolically very much alive (in the form of a ghost that haunts Hamlet). Both of these exceptional deaths are effects of predominant discourses and practices of death (what constitutes a proper burial or proper mourning at their particular historical and cultural moments). But current discourses and practices of death are only effects of the undead. The undead, therefore, is pure difference, something that does not exist in time but out of time, and which then erupts to reconfigure everything only to be naturalized, repressed, and regulated once again. The single most important task of established power is to regulate and manage (to destroy) the revolutionary nature of the undead. And prevailing discourses of life and death (in which life is ideologically separated from death) are the most effective means by which such destruction is attempted.

Whereas Žižek jumps from the undead within the imaginary of horror films to the undead as a philosophical and political force, Jean-Luc Nancy begins with his own body to make a similar jump. Here I am referring to how Nancy writes about the living dead in *L'Intrus* (1999)—the essay in which he presents a philosophically moving account of his own heart transplant eight years earlier. What I find fascinating in this piece is the depiction of the host-versus-graft disease, in which the host body must "somewhat" reject the donor organ. If there is no rejection, then one is effectively receiving one's own organ (whose death is the cause for the life-saving procedure in the first place), but if one totally rejects the donor's organ then one dies. Total rejection and total acceptance meet at a point of failure. Nancy is interested in how the stranger is internalized by the self, but at one point in the essay he makes an offhanded comment about how his son referred to him as the living dead—I find it hard not to read Nancy's essay as an argument for how life and death are overlaid with one another.

In the opening scene of *The Intruder*, Claire Denis's 2005 film adaptation of Nancy's essay, a young woman looks straight into the camera and asserts: "Your worst enemies are hiding inside, in the shadow, in your heart." Denis's reference to the heart is both figurative and literal, just as her film is both a radical departure from and a radical return to Nancy's original essay. And it is no surprise that the second shot of Denis's film is set right on the border-crossing separating Switzerland from France. The outside is already inside and the two sides that make up difference (two nations in this case) begin with simultaneity. This scene, in which a young female border guard is guiding her drug-sniffing dog through a stopped automobile driven by an older Arab man, is quite extraordinary. The tension in the scene is intense and fragile as the guard gives precise direction and encouragement to the dog. When something is discovered, the guard takes the dog to the side to reward it with petting and a treat. Denis then cuts to another scene without any reference to the smuggler and without any resolution to the scene's drama. Likewise, when the film's older protagonist, who has a bad heart, is riding his road bicycle quite strenuously up and down the hills of the rural valley, the tension is again ratcheted up. But as in the border-crossing scene, Denis does not resolve the drama. In these two scenes, and several others throughout the film, there is a formal representation of drama that is internal to the scenes themselves. The tension does not come from a prior situation, nor does the situation come from a prior tension, rather there is something simultaneous between the situation and the tension that turns out to be a cinematic enactment of the very philosophical problem Nancy is pursuing in his text

In *L'Intrus*, Nancy not only presents an account of his own heart transplant but the subsequent bone marrow transplant required to treat the lymphoma he acquired as a result of the original operation. We could say that Nancy's piece is a first-person account of these medical procedures if he did not put into question the very category of the "person" or "self" itself (the Heideggarian *mitsein*, or being-with, that Nancy retheorizes in *The Inoperative Community* and *Being Singular Plural*). The intruding organ enters as a disturbance, one that cannot be absorbed, neutralized, or denied its strangeness, a foreignness referring not only to the new heart but to Nancy's failing heart. As Nancy writes, "If my heart was giving up and going to drop me, to what degree was it an organ of 'mine,' my 'own?' . . . My heart was becoming my own foreigner—a stranger precisely because it was inside."[23]

Just as Nancy's body cannot repair the malfunctions of his own heart (to reprogram its all-too-short expiration date), his body cannot immediately absorb the shock of the newly strange organ. A successful transplant always turns on the success of treating the associated disturbances:

> The possibility of rejection establishes a strangeness that is two-fold: on the one hand, the foreignness of the grafted heart, which the host body identifies and attacks inasmuch as it is foreign; and, on the other, the foreignness of the state that the medical regimen produces in the host body to protect the graft against rejection. The treatments given to the one who has received the grafted organ lower his immunity so that his body will better tolerate the foreign element. Medical practice thus renders the graftee a stranger to himself: stranger, that is, to his immune system's identity—which is something like his physiological signature.[24]

One must give something away of oneself (in Nancy's case, both his failing heart and his immune system) and one must receive something from the other. But what one receives is not simply a healthy organ. Rather, it is the disturbance of the new organ, the disturbance that must be received or else the transplant will fail, for without the post-transplant havoc wreaked on the body there is no promise of an extended life. Without the disturbance, we are back to where we started, in need of something else.

Receiving the stranger must then also necessarily entail experiencing his intrusion. Most often, one does not wish to admit this: the theme of the *intrus*, in itself, intrudes on our moral correctness (and is even a remarkable example of the *politically correct*). Hence the theme of the intrus is inextricable from the truth of the stranger. Since moral correctness assumes that one receives the stranger by effacing his strangeness at the threshold, it would thus never have us receive him. But the stranger insists, and breaks in. This is what is not easy to receive, nor perhaps to conceive.[25]

If we substitute the word "stranger" for "death" in this quote, something interesting happens. We learn that death (like the donor organ) cannot be totally accepted (or else it is neutralized), nor can it be totally rejected (or else it is repressed). And this gets to the heart of the matter for Nancy, death is not the negative of life. Death and life do not function as a simple binary (or even dialectical) pair so that death functions to give meaning to life, and vice versa. Nancy writes, "'Death,' therefore, is not negativity, and language does not know or practice negativity (or logic). Negativity is the operation that wants to depose Being in order to make it be: the sacrifice, the absent object of desire, the eclipse of consciousness, alienation—and, as a result, it is never death or birth, but only the assumption of an infinite supposition."[26] Nancy sees it differently. Death does not suppose life, but dis-poses it. Nancy writes, "Dis-position is the same thing as supposition: in one sense, it is absolute antecedence, where the 'with' is always already given; in another sense, it does not 'underlie' or preexist the different positions; it is their simultaneity." Death is *with* life. Nancy as the living dead is Nancy living with death.[27]

For our purposes, both Žižek and Nancy are producing a third category (the undead and the living dead, respectively) in order to argue against the separation of life and death. But there are some key differences. For Žižek, the undead preexists discourses of life and death, but becomes fixed. Although for Nancy the living dead is "always already given" *and* it is a demonstrative, it shows itself in the midst of life and death—in their simultaneity. We have reached a critical difference between Nancy's deconstructive mode and Žižek's dialectical mode, a difference to be elaborated on when arriving at the other side of this rethinking of life and death, when the relative autonomy of life and death is retained. Before that elaboration, however, let's remain on the side of the problem that argues for integration by turning to Giorgio Agamben and his notion of "bare life," and then to Margaret Lock and her "twice dead."

Unsavable Life and the Twice Dead

As I mentioned in the final section of the global abyss, Agamben argues that bare life is relegated to one who is legally dead while biologically still alive. Referencing the human guinea pigs during the Second World War, Agamben writes, "Precisely because they were lacking almost all the rights and expectations that we customarily attribute to human existence, and yet were still biologically alive, they came to be situated in a limit zone between life and death, inside and outside, in which they were no longer anything but bare life."[28] This in-between life and death of *homo sacer* (sacred man), that person (with the concentration camp victim as the prototypical figure) who "may be killed and yet not sacrificed," serves an essential function in politics. The role of homo sacer is built right into political rule and sovereignty. Agamben writes, "The exception does not subtract itself from the rule; rather, the rule, suspending itself, gives rise to the excepting and, maintaining itself in relation to the exception, first constitutes itself as a rule."[29]

Foucault understood modern biopolitics as the threshold of the modern, as something that distinguished modernity from that which came before it, observing "the entry of phenomena peculiar to the life of the human species into the order of knowledge and power, into the sphere of political techniques."[30] In contrast, Agamben asserts the "idea of an inner solidarity between democracy and totalitarianism," so that biopolitics, in which life is submitted to the terror of death and is exploited as such to reproduce the power of sovereign rule, is the "originary" relation between law and life.[31] What does change over time, in Agamben's view, is the extent to which the exception becomes the rule. The exception of homo sacer becomes dominant. But this is not a linear process wherein the exception begins as a minor and occasional practice then over time gradually becomes more frequently invoked until it finally becomes dominant. Rather, the exception functions structurally for the rule—on a discrete temporal plane—so that it is always present to the same extent, and always simultaneous with the exercise of the rule. However, it can occasionally intrude or assert itself (a process distinct from "growing in strength" over time). For Agamben, this is precisely what has occurred in the contemporary global order of Guantánamo and contemporary medical practices that extend human life to a point beyond its own sustainability.

Agamben's *Homo Sacer* and *The State of Exception* are inextricably linked to his earlier work on the relation between language and death (in particular in *Language and Death: The Place of Negativity*). In *Language and Death*, Agamben begins with Heidegger's provocation about the essential relation of language to death, but wants to ground this relation in something other than negativity (as, he asserts, not only Heidegger but Hegel does—if not all of Western metaphysics). Agamben proposes to understand death not as the other side of life; he wants to appropriate the barrier that separates life and death. Agamben does not celebrate or wish for a return to the sacredness of life, biological life (*zoe*) as free from political life (*bios*), but life that is not subjected to the power of death, for it is precisely this subjection that leads to abandonment (to murder without sacrifice).

When Agamben argues for a life lived beyond the reach of the law (or what he calls a happy life in *Means without Ends*), he effectively argues for the end of death and life as we know it, as well as the end of the separation of life and death.[32] It is here that Agamben appeals to Walter Benjamin and his notion of the "saved night"—nature that is sufficient in itself and does not need to be saved by human beings—and turns it into a life that is saved by being unsavable, and it is only at this point that the dialectic between life and death comes to a "standstill."[33] (I am extrapolating this Benjaminian turn of phrase the way Agamben does when rejecting the separation of human and animal.) This standstill is not one that is instrumentalized by the state, as is the case when advanced medical technologies extend life to such a degree that (as in Agamben's example of Karen Quinlan) the body enters "a zone of indetermination in which the words 'life' and 'death' [have] lost their meaning, and which, at least in this sense, is not unlike the space of exception inhabited by bare life."[34] The question becomes how the indetermination can exist and not be exploited by the state. The state, in other words, has become successful at unsaving a life by saving it, which has now become the rule due to the prolific rise of contemporary medical technologies.

Near the end of *Homo Sacer*, Agamben writes the section titled "Politicizing Death," focusing on the changes brought by advances in medical technologies. He begins with the category of "overcoma," in which technologies sustain a stage of life beyond the termination of all vital bodily functions. Agamben realizes that such a state redefines death and takes on an even more significant dimension when put next to developments in transplant technologies. "The state of the overcomatose person was the ideal condition for the removal of organs, but an exact definition of the moment of death was required in order for the surgeon responsible for the transplant not to be liable for homicide."[35] The consequence of this is that death "becomes an epiphenomenon of transplant technology." For Agamben, this is a contemporary instantiation that attests to the paradigmatic force of bare life. Although appearing for the first time in the hospital room (and representing a new biopolitical threshold), this contemporary example is based on the same principle as the concentration camp. This leads Agamben to his final provocation: "In modern democracies it is possible to state in public what the Nazi biopoliticians did not dare to say."[36]

One immediate question for Agamben is where the so-called underdeveloped societies fit into all of this. We know that the majority of the world's population considers the fear of too much medical care a luxury. This leads to the problem of not only transnational class differences, but significant cultural differences that shape the biopolitical. And this takes us to Margaret Lock's notion of "twice dead," which confronts the way the current reinvention of death (due to advances in transplant technologies) not only complicates any fixed definition of death, but the way "twice dead" refers to how death might function in one culture while functioning quite differently in another.

As a medical anthropologist, Lock studies the differences between transplant cultures in Japan and in North America. It is well known that transplant culture in Japan is quite underdeveloped, however much the so-called brain-death problem is hotly contested. This leads Lock to wonder why the condition of brain death in Japan had not become legally recognized until 1997 and why organ transplantation is not considered an unequivocal good. Instead of essentializing the Japanese and making homogenized claims about cultural difference, Lock begins by turning the questions around. She asks, "Why did we in the 'West' accept the remaking of death by medical professionals with so little public discussion?"[37] Lock's point here is not to discount cultural differences, but to properly historicize them—which means resisting the urge to invoke ahistorical tradition. For example, the majority of Japanese who are skeptical of transplantation are skeptical not because of deep-seated Shinto or even Buddhist beliefs, but because of a very modern suspicion that government power is in cahoots with the medical industry. Despite the extent to which the gruesome medical experiments carried out on the Chinese or the state-sponsored cannibalism during the Second World War might be repressed from memory and suppressed from the nation's textbooks, a serious mistrust of government intervention with the human body persists for many Japanese.

Notwithstanding the resistance to reifying Japanese tradition, Lock does argue that the "majority of Japanese live and work with ontologies of death that differ from those in North America."[38] Japan is a particularly informative country to analyze in this respect because it is quite advanced in terms of medical technologies and is, comparatively, one of the richest nations in the world. Underdevelopment and a lack of expertise do not wash as explanations for why transplantation lags so far behind transplantation in the West. Lock argues that in the West a new uniform definition of death was easier to institute because of the priority granted to the individual. When the brain of the individual is no longer alive, then the individual is considered dead. But in Japan, where there is a different relationship between the individual and the group, the death of an individual's brain is not the chief criterion of death. Because the family and the social group play a central role in end-of-life care in Japan, death is something shared and not solely the domain of the dying person. Death, in this context, exceeds the dying individual.

This different priority given to the individual and the group relates to another key difference between Japan and the West, that of gifting. In Japan, gift giving is usually performed as a way to cultivate social relations among members of a group, especially as a way to negotiate power relations within a hierarchy. Requests are usually accompanied by a gift (that would be tantamount to a bribe in the West), a practice that expresses a true and candid recognition of how the system of power works in Japan. This also means that gifts are not given to strangers. But the "gift of life" that transplantation gives is almost always between strangers—the hallmark of the transplant exchange. The different "ontology of death" that Lock stresses is another way of arguing that life and death are malleable, especially the relation between life and death. Lock writes, "All along we have imagined that by clearly demarcating life and death, we would retain objectivity and integrity, ensuring that no one not yet dead could be counted as good-as-dead. The game is up, it seems, leaving us to face another unsavory discovery—that death is not amenable to our efforts at is mastery; it will not be pinned down once and for all."[39]

A few years ago, this lack of demarcation was actually institutionalized in Japan, when the University of Tokyo inaugurated a new academic program titled "Death and Life Studies," which brought together the school of medicine, the humanities, and the social sciences. The title of the program suggests a set of reversals away from the assumption of the linear progression from life to death, as well as away from the assumption that life and death can be separated into discrete objects of study.[40]

Death and Discontinuity

Žižek, Nancy, Agamben, and Lock all remove the line separating life and death, and each thinker gestures to a different politics out of this removal. But we can come at the relation between life and death from the other direction (one that stresses how death must retain its autonomy from life and thus maintain the gap that is the space of utopia). Here I am thinking about the work of Herbert Marcuse, Ernst Bloch, and Fredric Jameson. Only two years after publishing *Eros and Civilization*, Marcuse wrote an essay in 1957 titled "The Ideology of Death," wherein he argues that when dealing with death the Western philosophical tradition (beginning with Socrates and moving to Hegel and Heidegger) promotes two equally questionable ethics: the first is the stoic or skeptical acceptance of death, and even the repression of its inevitability; and the second is the "idealistic glorification of death" as that which gives meaning to life. Marcuse is particularly critical of what he calls the existential privilege granted to death: "From the beginning to the end, philosophy has exhibited this strange masochism—and sadism, for the exaltation of one's own death involved the exaltation of the death of others."[41] Marcuse is concerned with how the acceptance of death carries with it "the acceptance of the political order." Experiencing death as a metaphysical limit effectively limits the imagination (I view this metaphysical finiteness as a correlative to Marcuse's critique of scarcity in *One Dimensional Man*). This is where we get as close as Marcuse ever comes to thinking about death and utopia: "With death as the existential category," he writes, "life becomes earning a living—rather than living, a means which is an end in itself. The liberty and dignity of man is seen in the affirmation of his hopeless inadequacy, his eternal limitation. The metaphysics of finiteness thus falls in line with the taboo on unmitigated hope."[42]

This taboo on unmitigated hope brings us to the other great analysis of death and utopia, namely, the one written by Ernst Bloch in the third volume of *The Principle of Hope*. Bloch confronts death and utopia in a section titled "Self and Grave-lamp Or Images of Hope Against the Power of the Strongest Non-Utopia: Death." These eighty pages magisterially move from various religious traditions to what Bloch calls enlightenment and romantic euthanasias and the "secularized counter-moves" of nihilism and socialism. Finally Bloch arrives at the following: "Death is thus no longer the negation of utopia and its ranks of purpose but the opposite, the negation of that which does not belong to utopia in the world, it strikes it away . . . in the content of death itself there is then no longer any death but the revelation of gained life-content, core content."[43] And later in *The Principle of Hope*, when stressing the noncoincidence of life and death, Bloch writes: "The old saying of Epicurus that where man is death is not, and where death is man is not, comes true here."[44] Like Marcuse, it is by way of a rigorous materialism that Bloch resists the metaphysical and existential questions and leaves open the space for a future death that can be experienced in a way robbed "of its sting," as Jameson put it.[45] And here we see in Bloch a hint as to the progressive potential of the already dead. Death is recognized, inevitable, anticipated already from the present; however, it is not the death we currently know, the death that structures our current understanding and behavior, the death we prepare for with wills and insurance and existential dread. It is a death that cannot be imagined or represented, because it marks the termination of a life we are not yet living. Therefore it comes to us only vaguely, in flashes and unformalized.

When commenting on Bloch in *Marxism and Form*, Jameson writes, "Death in such a world, has nothing left to take; it cannot damage a life already fully realized."[46] But, as Jameson makes clear in *Archaeologies of the Future*, this utopia lies beyond our capacity to imagine it, and like death will always retain its absolute discontinuity. In *Valences of the Dialectic*, Jameson goes on to argue that discontinuity (and incommensurability) is part of a dialectical movement. He writes, "Yet this excess or inassimilability itself constitutes a dialectic—'between the non-dialectizable and the dialectizable'—which potentially renews the dynamics of the process and opens up the possibility of a new and enlarged dialectic in its turn, the clock of dialectical temporality once again beginning to tick."[47] With this in mind, we can argue both positions: that the inassimilable aspect of death (that aspect which is irreducible to life) is a radical opening of the system, an opening (as it is for Marcuse and Bloch) to a life that might only be possible in the future; and, along the lines of the first set of thinkers, that the absolute discontinuity of death marks a static outside which, therefore, characterizes a closed and nondynamic internal system.

This mention of the dialectic already suggests that it would not take too much work to turn all of this around and discover that the figures of the undead, the living dead, homo sacer, and the twice dead, as presented by Žižek, Nancy, Agamben, and Lock, hold within them the utopian space between life and death, while Marcuse, Bloch, and Jameson can be just as easily understood as fusing the two conditions. For example, we have already established that Žižek's undead functions differently in the three different orders (the Imaginary, the Symbolic, and the Real). His notion of pure difference is a difference that comes prior to the differentiated terms, despite the fact that there is a simultaneity between the way this difference functions in relation to the terms themselves (in this case, the terms of life and death), while there is still a radical separation between the differentiating force and the terms in a way that can only make sense when we respect the different logics of the three orders. We see this same unambivalent engagement with contradiction in the work of Nancy, when he argues that death does not suppose life but dis-poses it, in which this dis-position is at once absolute antecedence and not preexisting the different position. Likewise, Agamben's use of Benjamin's philosophy of history (which is the ultimate argument for discontinuity, with its "emergency breaks of history," "ajar doors" awaiting revolutionary events, and banal fragments exploding the continuity of homogeneous historical time) persists and breaks through his Heideggerian temporarily of being that is absolutely and inescapably contained and informed by its own history. And for Lock all we have to do to see this doubleness is to focus on the parts of her work where she emphasizes Japan's radical difference from the West while remaining adamant that such difference is not essential, unique, or stable. But all of these somersaults should not come as a surprise, since they are ultimately symptoms of the problem itself, of the antinomy of life and death.

We began this part on the already dead with the new desire to insure against death, in the forms of death insurance for individuals and death insurance for capitalism. Both events revealed the impossible double desire to separate death from life and to absolutely integrate them. Mainstream discourses about death and life invariably separate death from life so that death functions to terrorize life from beyond, forcing upon us both acquiescence and unbearable sacrifice. The other side of these mainstream discourses is the equally insidious discourses that inextricably link life with death, so that death acts to fundamentally limit the imagination and contain radical political desire. The first set of thinkers I explored negates the false separation of life and death, while the second set of thinkers negates the false sameness. Both groups, however, affirm that the relation between death and life is the crucible of modern ideologies.

The already dead both negates the reactionary separation and identity of death and life and affirms their radical separation and identity. In this way the concept is a double negative and a double positive. Death is not the negative of life; rather, death is the negative of death. This means that the negative of death is not its presumed opposite, namely, life, but the place in death where death exceeds itself and opens up to an alternative configuration. Likewise, life is the negative (the excess, the virtual) of life itself. But these negative valences of life and death exist as potentials, negating through their objective nonexistence (and unrepresentability) the insufficient existing categories of life and death. This doubleness requires a whole new set of relations (if not a whole new theory of relation itself), namely, the relation of death to itself and of life to itself.

But what does it mean to say that the already dead is also a double positive? It is here that we might remember the old university legend of a pedantic academic prattling on about the grammatical abomination of the double negative and then, as if possessed of some brilliant theoretical insight, asks his students if there is such a thing as a double positive. The professor luxuriates in the silence, until someone from the back of the room sarcastically yells out, "Yeah, yeah," before standing up and unceremoniously leaving the lecture to the horror of the professor and to the exhilarated disbelief of his fellow students. The only way to escape such an impossible and infuriating question is to crack open a new possibility in a different register—in the case of the student's response, we shift from the register of circumscribed reason to that of emancipated style. The student hit the bull's-eye by way of formal invention, despite the fact that the target did not exist within the realm of the dominant discourse. But the student's double positive has a certain singularity to it; it only works at a particular moment and is wholly contingent on the historical conjuncture it confronts. In other words, the "yeah yeah" changes all the rules, but once the rules are changed it loses its force and immediately becomes another element reinforcing a newly emerged antinomy. This illustrates the contemporary potential of the already dead. Just as revolutionary consciousness is snuffed out by the logic of the chronic (in which the possibilities to imagine and desire a radically different future are suppressed by the promise of management), the already dead flashes the radical possibility of usurping dominant discourses of life and death and reigniting revolutionary consciousness.

Revolutionary consciousness, however, does not necessarily lead to revolutionary social change. As we have observed in several different ways, knowledge (consciousness) does not necessarily change anything. This directs us to the following key questions: What is the relation among the political, psychological, cultural, and spiritual aspects of the already dead? And how might such integration challenge the actually existing political situation?

The most famous political invocation of the already dead in recent times appeared on the wall of a bank in San Cristóbal de las Casas on January 1, 1994. "Here we are, the dead of always, but now to live." This phrase was then used as a head note on the communiqué issued on January 6 by the general command of the Zapatista Army of National Liberation. Later that year, when explaining how the Zapatistas mustered the courage to fight not only the powerful Mexican state but the larger global neoliberal complex, Subcomandante Marcos said, "We have nothing left to lose. We are already dead."

In *The Tibetan Book of the Dead*, the single most important point communicated to the recently deceased is that they are already dead and that they are now in the *bardo*, a between space that exists after death and before rebirth. The reading of the book takes place over a period of forty-nine days and is a guide to the dead on how to move through the six transitional realms. Without the real-time performance of a lama reading the text, the dead would invariably forget that they are already dead and attach to the lower realm of infantile lusts and hungers. The book is not only about death, however, and the bardo does not only represent the gap between death and rebirth. Rather, the truth for the already dead is the truth for the living, and the bardo represents all gaps, most notably the gap between life and death in which the living always exist. It is only at the moment when the living remember that they are already dead that the possibility for liberation emerges.

This spiritual challenge to dominant discourses of life and death not only relates to the political force of the Zapatista consciousness, but to the psychological force of the psychoanalytic experience. The lama performs the ritual, reads the text aloud in the presence of the corpse. The dead can still hear and is influenced by and influences the performance. But the performance is for the living as well—most significantly to inspire them to see that their infantile desires are of their own making and can be overcome by recognizing that these desires, although very much human, are not inevitable. The lama, therefore, guides not only the dead but the living through a dynamic process that resembles the praxis of psychoanalysis. In fact, Carl Jung, who in the mid-1930s wrote the famous introduction to this text, has already made the point that the psychoanalytic relationship between analyst and analysand is the only Western practice that resembles the praxis of *The Tibetan Book of the Dead*.[48]

First written in 1935, Jung's "Psychological Commentary" is notable for many reasons and his attachment to the Buddhist text can help us elaborate the temporal dimension of the already dead. *The Tibetan Book of the Dead* coordinated with Jung's own understanding of psychoanalysis based on supratemporal archetypes—universal forms shared by all human beings. *The Tibetan Book of the Dead* articulates three main stages on the way to rebirth: the first is *chikhai bardo*, the spacious luminance the dead experiences right after dying; the next is *chonyid bardo*, a contemplative, transpersonal period in which the dead faces both wrathful and peaceful deities; the final stage before rebirth, *sidpa bardo*, is occupied by infantile hunger and lust. But Jung notably reverses this order and reads the text backward, beginning with the vulgar birth instincts of the sidpa and culminating in the imageless stage of luminance. Jung explains that this is the best way for a Western mind to understand the text, a way that follows a progressive trajectory from nastiest to finest. Throughout the commentary,

Jung fiercely criticizes Western science and philosophy, most notably for not being able to hold the truth of a contradiction. For example, Jung is impressed with how Buddhism can explain deities as human projections while at the same time understand them as real. It is this "both and," for Jung, that is inconceivable by the "either or" rationalism of the West. Why Jung does not pause to consider Hegel (in which appearance is not opposed to the real, but appearance is the real itself) or Marx (in which ideology is not false consciousness, but real consciousness itself) becomes clear when we realize that Jung's real target in the commentary is Freud.

Jung criticizes Freud for getting stuck in the cul-de-sac of infantile sexual instincts, which coordinates with the third stage (the sidpa bardo) of the book. Because Freud, according to Jung, only goes back to what is effectively the Oedipus complex, Freud can only have a negative valuation of the unconscious and a negative valuation of human kind, since we are destined to arrive and remain in this miserable realm. But Jung wants to move to the next stage, that which results from the psychic residue of previous existences and takes the individual from the personal and subjective to the "anytime and anywhere" categories of the imagination. Now we can grasp the real reason, however disowned, why Jung reverses the order of the text, putting the stage of supratemporal karmic illusions after the stage of birth instincts on the way to luminance—to self-servingly argue for why his own thought is a corrective step beyond Freud's.

At stake here is a theory of temporality. Moving backward through the text, Jung employs a teleological temporality in which there is a progressive development toward the goal. But the *Tibetan Book of the Dead* resists such a temporality. Rather, there is a movement through the stages of the forty-nine days, but at any moment a radical break from the linear development can occur. We can argue the same thing about Freud. Freud resists such a progressive (or retrogressive teleology) and holds open the space for a radical break from the dominant human habits of desire. And he does so by stressing the one temporal category that Jung invariably rejects: social history itself.

Jung's temporality allows for only two temporal categories, the subjective and personal time of individuals and the eternal and transpersonal time of archetypal forms. The first is absolutely particular and the second absolutely general, thus arriving at the ahistorical at both ends. Jung ascribes the particular to Freud's interest in sexuality and the general to his own understanding of archetypes. But history as a third category, in which a social formation (in our case, capitalism that is produced by humans) produces the limits and possibilities of human subjects (all the while itself transforming) takes us to the ground on which politics, culture, and psychology is most radical—radical in terms of how this ground inscribes its dominant forms into subjectivity (however differently) and how it is always open to revolutionary reconfigurations.

Freud is also writing about religion in the late twenties and early thirties and, like Jung, has a particular interest in Buddhism. The beginning of *Civilization and Its Discontents* begins with Freud's response to his friend Romain Rolland. Rolland agreed with Freud's previous work (*The Future of an Illusion*, written in 1927) on the function of religion to fulfill deep-seated human wishes, but admonishes Freud for not properly accounting for the "oceanic" and "eternal" experience of religious sentiment. Freud explains that such a feeling defies description in the first place and, therefore, all there is left to do is to try to explain it as a symptom, which Freud does by linking such a boundless, eternal feeling to primary narcissism, or to the longing for a lost wholeness that adults project back onto their pre-Oedipal pasts. Religious desire, therefore, is not unrelated to the workings of the pleasure principle, whose sole function is to reduce tension. But death is a "zero degree of tension," leading to the inextricable relation between life and death. In fact, when one of the first Japanese scholars of psychoanalysis, Yabe Yaekichi, traveled to Vienna in 1932 to meet with Freud, Freud asked why *Beyond the Pleasure Principle* (the most important Freudian text to elaborate the death drive) was one of only three of his texts translated into Japanese. Yabe explained that the relation between life and death as worked out in that text was similar to the relation between life and death within Buddhism (not eschatologically, but dialectically) and was, as a result, most accessible to the Japanese reader.[49]

For Freud, eros distracts us from death and, therefore, is the most potent weapon societies use to placate the potentially revolutionary energy of its subjects. Eros is recruited by civilization to subdue its populations. The injunction to "love they neighbor" is effectively a strategy by established power to force subjects to repress their instincts, to sacrifice freedom, which in turn produces guilt, the mollifying condition of modern life par excellence. Freud sees this process as always occurring, with every social formation forcing its subjects to renounce its libratory instincts. But each society does not do this in the same way—and this is the key to what we might consider Freud's historical materialism. His comments about communism are especially important in this regard. Freud is sympathetic with the communist desire for equality and even for the abolition of private property, but he is convinced that even after such revolutionary measures aggression will still exist, resulting in the need to repress a population with all of the associated guilt that such a process engenders. "Aggressiveness was not created by property . . . One only wonders, with concern, what the Soviets will do after they have wiped out the Bourgeois."[50] The point here is that there will be qualitative differences in the way different societies manage their populations. There will always be aggression, and its social control will necessarily produce many guilty subjects. But the form of this aggression (and even the form of this guilt) will be different, thus leading to different configurations of social power—configurations that can be evaluated as more or less just. Freud's historicization of the instincts, therefore, opens the way to a potential political solution (to a revolution of the actually existing conditions), while Jung's opens the way to a potential religious solution (to a search beyond history and to a deemphasis of the actually existing conditions).

In *Eros and Civilization*, Marcuse elaborates this historical dimension of Freud's *Civilization and Its Discontents*: "The unhistorical character of the Freudian concepts contain elements of their opposite."[51] But Marcuse does want to leave open the space for a nonrepressive society. He sees Freud's understanding of eros as limited by a capitalist logic. In other words, Freud understands eros as a scarce resource that each civilization hoards as a way to repress its population. Marcuse criticizes Freud for transferring what is essentially a capitalist ideology of scarcity in terms of resources (an ideology that is effectively mobilized to justify inequality) into a transhistorical psychological principle. Since the discourse of scarcity is peculiar to specific modes of production, there does not have to be a scarcity of eros. For Marcuse, the reappropriation of eros is the ultimate political act. He wrote, "Today, the fight for life, the fight for eros, is the political fight."[52]

Marcuse wrote about this fight in 1966 in the political preface to *Eros and Civilization*, eleven years after the book's original publication. "Today, however, the fight for death is the political fight."[53] This claim returns us to the new chronic and to the already dead. The new chronic is the most recent way to steal our deaths from us. We either die with no care (for which there is no death) or we die with total management (for which there is no death). Or we reclaim time and reclaim death in the mode of the already dead, thereby opening up new deaths, and new ways to die. And opening new ways to die means opening new modes to live.

The already dead have been killed, but have yet to die. The already dead, however, are not just dead symbolically and alive biologically (like Antigone) or dead biologically and alive symbolically (like Hamlet's father); rather, the already dead are dead biologically and alive biologically, as well as dead symbolically and alive symbolically. Through formerly unthinkable interventions in the context of global capitalism that produce an alternative temporality, the already dead live an impossible life within an inescapable death. Amid all the misery, this impossible space also amounts to a free zone in which the already dead can transgress the structural limits of the present situation. It also gestures negatively (by its very impossibility) toward radical structural reconfiguration.

There is a term within fiction called "comic book death," referring to a character who has been killed off but subsequently returns. In the serialized graphic novel, the hero or villain returns to sustain the narrative. Comic book deaths have also been famously employed by the likes of Sir Arthur Conan Doyle, who killed Sherlock Holmes in the story "The Final Problem" (1893) only to resurrect him ten years later in "The Adventure of the Final Problem." The writers of the TV series *Dallas* made use of the same trope, creating a character who dies in one season only to return in the next, after it is revealed that everything that happened in the previous season was a dream. In the 2007 film *Michael Clayton*, Clayton unexpectedly steps out of his car and walks up a hill to look at some wild horses. The serene scene is then violently upset as a bomb blows up Clayton's car. Immediately realizing that the bomb was meant for him, Clayton returns to the burning car to throw his watch and wallet into the flames. He then runs off to figure out his next move. At this point, the would-be assassins believe Clayton is dead, providing him with a space of freedom to act. There are numerous examples of Hollywood films in which a character on the run takes advantage of his presumed death. Insurance windfalls, new love affairs, and revenge become possible as the mourners mourn and the assassins collect their fees.

For the comic book death to work, however, the dead must claim their deaths. They must not deny death, but identify with it and, one step further, reconfigure their own identities by means of it. This, in turn, is what upsets the law and allows for something new to happen. Holmes, for example, travels the world without letting Watson know that he is alive. In order to remain undiscovered, the character must work in the shadows and by himself. If the character does risk reaching out to someone, it is invariably to a single friend who, after first overcoming the trauma of the false death, must guard this new secret with his or her life. This is one of the two fundamental ideological functions of the comic book death as it is employed within popular culture: the dead protagonist must act on his or her own. In the end, this is a narrative strategy that works to reject, denounce, and deny collective action. If the character joins with others then he compromises his freedom.

The other ideological function has to do with time. Comic book deaths are also referred to as "retcon" narratives, as in "retroactive continuity." In order to bring back a character from the dead, the author must go back in time to expose loopholes that allowed the protagonist to escape what everyone believed to be certain death. Dreams, medical miracles, or terrific luck usually enable a retcon, which, in the end, reasonably explains away the impossibility of a dead character not being dead. Chronic time is reestablished, closing off the freedom (for the author, reader, and characters) that the suspended death permitted. While the first ideological function of these narratives undermines new forms of collectivization, this second function snuffs out alternative forms of temporality. Today, the already dead, like the comic book dead, must not only identify with their deaths, but, and as they do not in the conservative recuperation within the fictional narratives mentioned above, they must also find ways to join with others in a collective project and resist the new chronic time of global capitalism.

A Nonmoralizing Critique of Capitalism

Death offers an ideal portal into these projects, beginning by enabling a nonmoralizing critique of capitalism. To explain this connection, let's first establish what a nonmoralizing critique is. First, it is not personally motivated. Of course, every action is personally motivated insofar as it comes from an individual person and is necessarily fashioned by conscious and unconscious desire. In this case a nonpersonal critique of capitalism means that one first recognizes that one is necessarily part of capitalism, necessarily wrapped up in its ideologies, and that one shares this necessity with others, both friends and enemies. There is no escaping capitalism, since capitalism is not only the production and consumption of commodities, but a certain mode of production with special forms of exchange, meaning-making, social relations, desire, communication, and thought that necessarily insinuate themselves into our very beings, so much so that attempting to avoid them is like trying to avoid our deepest habits, from the way we hold our bodies to the way we think about how we hold our bodies. This inextricable relation to capitalism (which affects the very way we understand and represent it) leads to the recognition that any critique of capitalism is necessarily social, necessarily part of something that exceeds the individual producing the critique.

Second, this nonmoralizing critique is not personally directed. The critique, rather, is directed toward the structure, system, and logic of capitalism, which requires less a scathing rhetoric against individuals and more an analytic understanding of how capitalism works. The capitalist system works to produce greedy and corrupt capitalists (ones who certainly deserve condemnation), but to begin with a criticism of them is counterproductive—not only because the dominant system of representation (media, mass culture, pedagogy) is based on a sophisticated defense of these very individuals and their practices (so that to engage in a shouting match in the contemporary mediascape is to risk neutralizing all critique), but because to go after the successful capitalists undermines the analytical skills required to understand the larger system. Capitalism is a tremendously complex system, which was proved once again during the financial meltdown of 2008, when the derivative schemes were so intricate that the only people who were capable of dismantling them were the very individuals who invented them in the first place.

Finally, to direct a critique at the system and not at the individuals who manage and defend it is to reaffirm a belief in the system itself, in the system as such. When I argue throughout this book that crisis occurs in capitalism not because capitalism has gone wrong but because it has gone right, I am arguing that there is a certain cause-and-effect logic that can explain such events as war, poverty, and illness (these effects are products of other systems as well, but the specific configuration of war, poverty, and illness within capitalism is qualitatively different than their configuration within different systems). Without the recognition of a greater logic special to each system, one effectively abandons politics as such. A nonmoralizing critique of capitalism wants to reaffirm a belief not in "the system" (and certainly not in the capitalist system), but in the "system as such." Keeping the problem of the system in the foreground (by deemphasizing a moralizing critique, for example) enables a consciousness of the historical fact that capitalism is a system that came into being at a moment in history and will go out of being in the future. Without this belief in the system of capitalism and, more importantly, in the very reality of the system, revolutionary politics is impossible.

This leads to the third criterion of a nonmoralizing critique of capitalism: since there is always something within a system that escapes the systemic logic, something any critique cannot fully incorporate, one must be open to—and try to hold—the contradictions of capitalism, rather than immediately manage, resolve, or repress them. This is to say that capitalism can produce magnificent qualities while still causing heartbreaking destruction. To recognize this is also to recognize the history of capitalism, especially the unquestionable liberating effects that its founding revolution enabled. This simple fact sustains a nonmoralizing critique, since it denaturalizes capitalism, opening up a comparative analysis with other social formations.

This comparative analysis (which also means comparing capitalism to other formations that do not yet exist) is based not on the ideological claims and desires of different systems (democracy and freedom, for example), but on what each system delivers, such as health care, a healthy natural environment, opportunities to experience diverse pleasures, social equality, individual justice, nourishing food, and secure shelter. A nonmoralizing critique, therefore, prioritizes outcomes and remains unconvinced by nonsocial and ahistorical justifications and arguments, such as the complacent recourse to the scarcity of natural resources or the inherent greediness and goodness of human beings. This comparative impulse also inspires formal experiments with alterity, from social modeling to science-fiction narratives. Such exercises themselves should not be justified by any moralizing critique, but neither should they be discouraged by the constraints of practicality or impossibility. To make the impossible might very well be impossible, but the very act of imagining it can change the realm of possibility.

This is another way to make sense of the relation between life and death, as well as the already dead. It is logically impossible to claim simultaneously that "I reject the regime of management and demand my right to die" and "Death does not exist." However, what this impossible double-claim produces is a destabilization of the present, a death of the sick body in crisis (currently being managed) that is at the same time a radical affirmation of life (the life of another body whose death is nothing like the death we currently imagine). The already dead are those who inhabit the space of this contradiction while unrelentingly insisting that it cannot be resolved within current structural formations. The already dead refuse, thus, either to die or to be alive until these categories can be remade to accommodate the unique and new existence the already dead experience.

This leads to the final criterion for a nonmoralizing critique of capitalism: if one appeals to evil or righteousness, then these qualities and acts should be understood as symptoms, rather than causes, of the very system under question. Evil acts do not cause capitalism's crises and then recuperate these crises by dispossessing individuals of their wealth and dignity. This process of crisis and dispossession is built right into the system itself and, as in any machine, can do certain things but not others. Instead of anthropomorphizing capitalism with histrionic claims of how evil or righteous it is, a nonmoralizing critique sees it for what it is—a human-built machine that performs various functions based on specific rules and fundamental principles. Such a critique would generate a certain degree of respect for capitalism based on how capable it is at performing such tasks— even if such tasks are as brutally cruel as allowing millions to die of treatable illnesses. Instead of incredulity and counterproductive anger, a nonmoralizing critique generates a clear voice (however angry) and a measured response (however poetic) that does not retreat from the most painful and beautiful aspects of everyday capitalist life.

Now the question becomes how death and the already dead relate to these criteria of a nonmoralizing critique of capitalism. First, there is the role of crisis. Crisis transforms for the already dead from the short term to the long term, in which one is still in dire straits while the immediacy of the danger, has been deferred. In this way, crisis is not something that comes and goes, but is always present. Once diagnosed, one will always have cancer and, therefore, will always live with the threat of its acceleration or relapse. The already dead are exposed to a logic of crisis, which is homologous to how crisis functions within capitalism. This also requires a form of awareness that lends itself to an astute economic analysis: one cannot forget that one has cancer, HIV, or a formerly terminal disease, however much life is normalized, just as one cannot forget that capitalism is always in crisis, however much day-to-day political-economic life is normalized.

But just as the ideologies of management do not work, neither do the ideologies of cure. Management and cure are usually employed to distract us from experiencing illness in a way that opens up to powerful personal and social questions. It is not as if the practical reality of management and cure are not important (of course, the new medications that maintain remissions and the potential for permanent cures are humbling and awe-inspiring), but the way discourses of management and cure occupy the experience of illness and the practice of medicine foreclose a variety of other experiences and practices.

The continual marginalization of palliative medicine is just one example of this, while the marginalization of alternative medicine is another. It is true that although so-called alternative medical practices might challenge dominant maintenance and curative discourses, they also work in perfect coordination with them. And here I am not only thinking of the way traditional Chinese medicine has recently lent itself to commodification and is seamlessly incorporated into the mainstream insurance plans of many Western health care systems (not to mention incorporated into a highly profitable global medical market). I am referring to how many contemporary discourses of alternative medicine work in coordination with the dominant discourses of Western medicine. For instance, despite the different logic of traditional Chinese medicine (which prioritizes a holistic model of interrelated functions, rather than an anatomical model that divides the physical body into separate symptoms and parts) there is a similar desire for maintenance and a cure that often functions to foreclose a wider analysis that links the medical with the political, economic, social, psychological, and cultural. This being acknowledged, the recent popularity of alternative medicine around the world is a symptom of a larger dissatisfaction with Western capitalist medicine. But this dissatisfaction will need to be combined with a more transdisciplinary analysis of medicine in order to grasp the crucial point that although many developed nations have public health care systems, the dominant paradigm of care, driven by for-profit pharmaceutical, biotech, and medical-device companies, is thoroughly privatized.

It is this wider, transdisciplinary analysis that opens up as dominant discourses of maintenance and cure weaken. The truth of the system reveals itself precisely when one is no longer colonized by the system itself. This clarity about the interrelationship of dominant discourses, however, usually comes too late for the dying, as the glimpsed conjuncture of how things work is extinguished by their imminent death. But with this imminent death deferred for the already dead, a new time of action emerges. This action, like the nonmoralizing critique of capitalism, is most significantly motivated when stressing the social dimension of death. Since one is already dead, the fear of death (that fear most powerfully exploited by the state) works differently and opens up a relationship to death and dying that is not wholly contained by personal struggle. Likewise, the already dead's struggle is not personally directed. This leads away from a focus on the miracle-worker doctor or cold-hearted hospital executive to a focus on the larger system in which doctors and hospitals operate. And when one's energies are no longer solely invested in specific individuals, the larger structure of power comes into the crosshairs of awareness.

But what does one do with this awareness? Make a revolution, since awareness changes nothing. But only collective action can bring a revolution about, which leads to the ultimate question: How will the already dead transform into a collective political subject? If their vulnerability does not prevent the required collectivization, then why would their national, racial, gender, religious, and class differences not do so? This in-fighting is precisely what capitalism is so successful at mobilizing. The already dead inhabit the two fundamental principles instrumental to radical change: collective action and the reappropriation of time. The category of the already dead, in other words, is precisely that subjectivity that effectively resists the logic of the chronic and freefalls in the global abyss. The already dead, however, do not constitute a political movement in the traditional sense. They, rather, evince a political consciousness that can inspire and inform political movements. The already dead already inhabit revolution— that is, a revolutionary consciousness informed by a certain way of living in time and with the future.

Over the past few years, a new phenomenon has occurred in cities around the world. People make up like zombies and walk together through the streets, stumbling crablike with moans of "brains" and other half-human utterances stolen from their favorite films. It is as if the movie theaters (with their countless zombie sequels) and the televisions and computers (with limitless zombie torrents and blog communities) could not contain the desire of so many zombie fans. They have taken to the streets, their gatherings resembling flash mobs, political marches, performance art, and simple afterschool parties of unreflective fun. Why the fans side with the zombies rather than the zombie killers, with the nameless film extras rather than the famous stars, must have something to do, no doubt, with the privileging of the zombie's irrepressible drive and unrequited hunger over the less attractive human qualities of neurosis and hysteria.

It is almost too easy to allegorize all this in an analysis in which the zombiewalkers express their resistance to everything from heteronormativity (their resistance to the straight couple in zombie films who invariably falls in love, or back in love) to the false overcoming of racial and class differences, all of which come into focus when confronting the zombies— the absolute enemy who must be annihilated before secondary concerns, such as justice or equality, can be faced. Perhaps the greatest resistance to be allegorized from all of this zombie fun, however, would be a resistance to the human itself—to the human as separate from not only animals and other living creatures but from the planet. Such an allegory reintroduces an ecological dimension that many might say was there from the start. And the allegory itself can be updated and recontextualized from decade to decade. Just a cursory glance at George Romero's filmography confirms this point, as the *Night of the Living Dead* (1968) is perfectly coordinated with the assassination of Martin Luther King Jr. and the civil rights, peace, and women's movements. *Dawn of the Dead*, made ten years later in 1978, takes us to the comforts of the shopping mall and to the seemingly conflict-free consumerism of the post–Vietnam War era. In 1985 *Day of the Dead* tells the story of survivors who must now deal not only with zombies but with a secret Reaganesque military cabal and its scientific collaborators. In *Land of the Dead*, made in 2005, class differences return with a vengeance as the rich are safely ensconced in a skyscraper while the poor fend for themselves against the now somewhat more sympathetic zombies.

But it is the problem of allegory itself that seems to need updating. Political allegory can be understood as the expression in a cultural register of something that cannot be articulated in the dominant political voice of the day. The switch of registers, therefore, is required because the contradiction at hand cannot be directly engaged, or at least when it is engaged it is falsely resolved in a way that reproduces the very system that generated the contradiction in the first place. When the object being allegorized is no longer unrepresentable, however, allegory turns tedious and becomes part of the dominant discourse that functions to repress a newly unrepresentable problem. The new vulgarity and transparency of capitalism leaves little to be allegorized, so that allegory seems superfluous; however, this situation itself promotes a false idea that capitalism contains no further secrets, that all has been brought to light. In fact, the cultural question today is what form (if not allegory) can reveal the current secrets of the system, those repressed internal crises that cannot come to the surface.

On its most basic level, the zombie film allegorizes how the collective of the modern nation (with the United States as the paradigmatic case), in order to sustain itself and manage its own contradictions, required a homicidal other to fight against. And the real horror is that the collective itself produces this enemy, by the very social system that brought the collective into being. The unrepresentable, therefore, is that one cannot have the modern nation without violence, which is in radical contrast to the nation's own fundamental narrative that understands its resort to violence as a response to the transgressive acts of others. Film was particularly well suited to capture this zombie allegory as its own experience was first organized around a collective of theater viewers (strangers to each other) that needed to repress the fundamental ideology of cinema, that cinematic meaning is made by our own desires and fears rather than by the film object itself.

But today this allegory has been played out—both the zombie film as allegory and allegory itself, which is not to say that there are no more allegories, but only that the function of allegory today has qualitatively transformed. The fact that millions of people die due to treatable illness and that refugee camps, illegal prisons, and labor exploitation abound is clear for everyone to see. And given that the excuses for this misery are explained primarily with concrete economic justifications (sustainability, deficit reduction) rather than grandiose political justifications (freedom, democracy) only goes to show that a new visibility has emerged that necessarily shifts the work of representation. Do we really need a zombie to shock us into recognizing that we are killing ourselves or that we feel exhilarated when we take a crowbar to the head of something that wants to destroy us? A new visibility (of an old capitalist reality) is coming into being that will shock almost no one. People the world over seem to know exactly how capitalism works and what it produces, which does not mean that we have arrived at the end of ideology (or the end of allegory, for that matter) but only that today the dominant ideology is the truth of the capitalist system itself. And, as we all know, truth is not easy to take.

The zombiewalkers seem to get this when they dress up as businessmen, housewives, and factory workers with very little makeup. They just walk down the street at a slow pace for the sheer act of being together. And the participants cannot help but look like a pack of zombies, who resemble more and more our everyday neighbors and coworkers (as both Kurosawa Kiyoshi's family drama *Tokyo Sonata* and Edgar Wright's comedy *Shaun of the Dead* reveal with deadly accuracy). For the zombiewalkers it is the performance itself, and the time of the performance, that is most interesting.

It seems, therefore, that without the ketchup smeared on their faces, it is harder and harder to distinguish between a zombiewalker and an average commuter. So here, as in ideology and allegory, the space separating the performance from the thing itself has been shortened to the point at which they almost fuse. But there is a profound difference between the zombiewalkers and the zombies they portray: the way they feed and the politics inscribed in this essential and unavoidable act. Zombies move through the city in search of food that comes in the form of human flesh and organs. The zombiewalkers walk until dinnertime. Perhaps one of the attractions to the fictional zombies is that they do not repress their fundamental need and, therefore, do not need to construct elaborate ideologies to mediate this need. We might go as far to argue that the attraction to the zombie is at the same time an attraction, however unconscious, to ideology critique.

The underlying question for any political movement is how the people will be fed, and how this feeding (the production, distribution, and consumption of food) will coordinate with the movement's values and principles. For the zombies, food is a public good. It is quite literally something that one takes from the very bodies of others and that, therefore, cannot be privatized, hoarded, or securely owned. In a distorted a way, the desire for zombies represents an alternative desire that dreams of something other than capitalism. But those desiring this something else cannot figure out how to feed without returning to the capitalist trough.

Most zombiewalks begin during the daytime. Perhaps so the walkers can be looked at in the natural light and then "feed" at night. But it is precisely this separation of day and night, and this desire for others to look at them, that turns the movement into a false allegory and compromises the radical promise of its political action. In contrast, it is precisely between day and night, human and nonhuman, allegory and the object being allegorized, cure and management, the chronic and the terminal, and death as continuous with life and death as absolutely discontinuous with life that we find the true force of the already dead.

The already dead do not desire to be who they are: they become, against their will, the already dead. The already dead learn how to let go while still holding on. They let go of the fear of dying. And they reclaim the fear for themselves. Death and dying might always be terrifying, but it is a terror that does not need to be defined by the state. The already dead also let go of the hysterical desire for salvation and cure. And they reclaim it. The desire for redemption, for reparation, for the impossible cure, might very well be our fatal flaw (that desire leading us back to prison), but to let go of these desires seems to open up the very possibility of their coming into being. This riddle, however, cannot be outsmarted in advance, but can only be solved without knowing it, in an entirely different realm that most closely resembles the utopia of praxis. It is for this reason that the already dead should not be fetishized or celebrated, but only recognized as a profound flash—and thus the possibility—of a radical future that already is.

Notes to the Introduction

1. Virilio, *The Vision Machine*, 66–67.
2. Schur, *Freud*, 484.
3. In a book tracking the swift rise of psychotropic medications in France and the United States and the current role of psychoanalytic treatments, Élisabeth Roudinesco quotes the French inventor of a well-known antidepressant drug: "Why is one happy to have psychotropic drugs? Because the society we live in is intolerable. . . . I am sometimes reproached for having invented the chemical straitjacket. . . . Without psychotropic drugs, there might perhaps have been a revolution in human consciousness, saying: 'We can't bear it any longer!' whereas we have continued to bear it thanks to psychotropic drugs." *Why Psychoanalysis*, 12.

Notes to Part 1: The New Chronic

1. Žižek, *The Sublime Object of Ideology*, 6.
2. U.S. Preventive Services Task Force, "Screening for Prostate Cancer: U.S. Preventive Services Task Force Recommendation," *Annals of Internal Medicine* 149, no. 3 (August 5, 2008): 185–91. "The USPSTF found convincing evidence that treatment for prostate cancer detected by screening causes moderate-to-substantial harms, such as erectile dysfunction, urinary incontinence, bowel dysfunction, and death. These harms are especially important because some men with prostate cancer who are treated would never have developed symptoms related to cancer during their lifetime. There is also adequate evidence that the screening process produces at least small harms, including pain and discomfort associated with prostate biopsy and psychological effects of false-positive test results."

3. See, for example, Margaret Everett, "The Social Life of Genes: Privacy, Property, and the New Genetics," in *Information Ethics: Privacy, Property and Power*, ed. Adam D. Moore (Seattle: University of Washington Press, 2005), 226–50.

4. Andrew Pollack reports the following: "But Pfizer is counting on cancer to help save the company. It hopes to reach $11 billion in sales of cancer drugs by 2018. That would be more than four times the category's sales last year of $2.5 billion, which represented only 5 percent of Pfizer's revenue. Cancer was once unattractive for big pharmaceutical companies like Pfizer. There were relatively few patients with any one type of cancer, and they died fairly quickly. By contrast, there were millions of patients with chronic diseases like hypertension who would take drugs for life. Cancer has become Novartis' second most profitable disease to treat. Indeed, the three main cancer drugs Pfizer now sells came to it with its 2003 acquisition of a rival, Pharmacia, a deal done mainly to acquire the arthritis drug Celebrex. But there are now many good cardiovascular drugs. Lipitor, the world's best-selling drug, will lose patent protection in 2011, and Pfizer failed to develop a successor. So Pfizer is scaling back cardiovascular research and has made cancer drugs one of its six focus areas. About 20 percent of Pfizer's more than $7 billion in research and development spending is on cancer, and 22 of the roughly 100 drugs in clinical trials are cancer drugs." Andrew Pollack, "For Profit, Industry Seeks Cancer Drugs," *New York Times*, September 1, 2009.

5. Bridgepoint distributed this advertisement in Toronto on bus shelters, billboards, and newspapers. Joanne Laucius, "Toronto Hospital Fights 'chronicitis,'" *Ottawa Citizen*, November 5, 2007.

6. "Complex Chronic Disease—The most pressing health issue of the 21st Century," http://www.bridgepointhealth.ca/researchthemes. "2007 Report on Ontario's Health System," *Qmonitor: Ontario Health Quality Council*, Toronto, 77.

7. Ibid.

8. Le Fanu, *The Rise and Fall of Modern Medicine*, 216.

9. William Bains and Chris Evans, *The Business of Biotechnology*, Cambridge University Press, http://www.fathom.com/course/21701784/session2.html. This web seminar is extracted from chapter 12 of Tatledge and Kristiansen, *Basic Biotechnology*.

10. Marcia Angell, "The Truth About the Drug Companies," in Law, *Big Pharma*, 14.

11. Law, *Big Pharma*, 14.

12. In May 2000, executives at the pharmaceutical corporation Merck chose not to pursue a study focusing on the cardiovascular risks of the painkiller Vioxx. Despite clinical evidence by outside scientists suggesting serious risks, Merck insisted that Vioxx was safe unless proven otherwise. Alex Berenson, Gardiner Harris, Barry Meier, and Andrew Pollack, "Despite Warnings, Drug Giant Took Long Path to Vioxx Recall," *New York Times*, November 14, 2004.

13. Law, *Big Pharma*, 69–70.

14. Cousins, *Anatomy of an Illness as Perceived by the Patient*, 29–78.

15. Christie Aschwanden, "Experts Question Placebo Pill for Children," *New York Times*, May, 27 2008.

16. Ibid.

17. Jopling, *Talking Cures and Placebo Effects*, 16.
18. Ibid., xix.
19. Ibid., xvii.
20. Ramin Mojtabai and Mark Olfson, "National Trends in Psychotherapy by Office-Based Psychiatrists," *Archive of General Psychiatry* 65, no. 8 (August 2008).
21. Freud and Breuer, *Studies in Hysteria*, 306.
22. Jacques Lacan, "Intervention," in G. Favez, "Le rendez-vous avec le psychanalyste," *La Psychanalyse 4* (1958): 305–14.
23. Freud, *Wild Analysis*, 3–9.
24. Ibid., 7.
25. Ibid.
26. Bruce Fink also refers to this joke as a way to distinguish the assumptions contained in it from the assumptions of Lacanian psychoanalysis. Fink writes, "Although the patient may initially claim to want to be relieved of his or her symptoms, he or she is ultimately committed to not rocking the boat." *A Clinical Introduction to Lacanian Psychoanalysis*, 3.
27. Freud, *Wild Analysis*, 7.
28. Ibid., 176.
29. Based on Oxford World's Classics translation, translation by Joyce Crick—Index of Dreams, 441–43. Freud writes in the foreword: "The only choice I had was between my own dreams and those of my patients under psychoanalytic treatment.... Reporting my own dreams, however, turned out to be inextricably tied to revealing more of the intimacies of my psychical life than I could wish or than usually falls to the task of an author who is not a poet, but a scientist. This was painful and embarrassing, but unavoidable; I have bowed to it then, so that I should not entirely do without presenting the evidence for my psychological conclusions." *Interpretation of Dreams* (Oxford: Oxford University Press, 2008), 5.
30. Sigmund Freud to Wilhelm Fliess, July 7, 1897, in *The Complete Letters of Sigmund Freud to Wilhelm Fliess, 1887–1904*, 254–56.
31. Ibid., 259.
32. Marx, *Capital*, 125.
33. Ibid., 104.
34. I elaborate on the different principles of a nonmoralizing critique of capitalism in part three of this book.
35. Uno, *Marukusu keizaigaku genriron no kenkyu* [Research on Marxist Economic Principles].
36. This is why many analysts request that their patients do not formally study psychoanalytic theory while in analysis. This was also one of the inspirations for Lacan to eliminate any difference between a training analysis and a so-called regular analysis. The merging of theory and practice can never be perfect. Each, rather, is always in tension with the other. Roudinesco, *Jacques Lacan*, 201–2; Schneiderman, *Lacan*, 87–88.
37. In 2008, I spoke to approximately twenty practicing psychoanalysts about Haneke's film *Caché* (2005). First, we watched the film together, and then I presented some ideas on how to think about the film in terms of psychoanalysis, not how psycho-

analytical themes are represented in the film but how psychoanalysis and film share similar theoretical and practical problems. As the discussion repeatedly fell back into representations of psychoanalytical problems on the level of the film's content, it became apparent that it was nearly impossible within the discussion to move to another mode of analysis. In fact, it is this impossibility that is foregrounded in the film itself.

38. Dienst, *Still Life in Real Time*, ix, 161.

39. Ibid., 164.

40. Hanks et al., *Oxford Textbook of Palliative Care.*

41. Ibid., 3. "Palliative medicine is the study and management of patients with active, progressive, far-advanced disease for whom the prognosis is limited and the focus of care is the quality of life."

42. Ibid., 4.

43. Ibid., 6. With the rise of managed care in the United States in the late 1980s, many medical schools and teaching hospitals suffered a severe lack of support for educational and research programs (either from government or through compensation for clinical services). In *Time to Heal*, Kenneth Ludmerer understands this shift in medical education in terms of larger shifts that affected the university, beginning in the post–Second World War moment with the instrumentalization of the university—with what Clark Kerr, in his widely read book *The Uses of the University*, called the multiversity, the professionalization of the university that led to its full-blown commercialization in the 1980s. Ludmerer writes, "As medical schools accommodated a diverse array of important activities, medical students became their forgotten members." Ludmerer, *Time to Heal*, 196.

44. As for palliative care experiments, I am thinking about how psychoanalysis may be practiced on the dying—a process that requires a reevaluation of both the time and purpose of psychoanalysis, as well as a reevaluation of how such a practice might be justified economically. I thank Dr. Gary Rodin for speaking to me about these experiments as they are practiced at Princess Margaret Hospital in Toronto.

45. Fredric Jameson, "Actually Existing Marxism," *Polygraph*: *Marxism Beyond Marxism?* 6–7 (1994): 170–95. An extended version of this article was published as a chapter in Jameson's *Valences of the Dialectic*, 367–409.

46. "But it seems paradoxical to celebrate the death of Marxism in the same breath with which you greet the ultimate triumph of capitalism. For Marxism is the very science of capitalism, its epistemological vocation lies in its unmatched capacity to describe capitalism's historical originality, whose fundamental structural contradictions endow it with its political and its prophetic vocation, which can scarcely be distinguished from the analytic ones." Jameson, "Actually Existing Marxism," 195.

47. This story is repeated not just by thinkers such as David Harvey or Giovanni Arrighi, but also in the Deleuze-inspired immanent Marxism of Michael Hardt and Antonio Negri, the Derrida-inspired postcolonial Marxism of Gayatri Spivak, the Lacan-inspired psychoanalytic Marxism of Slavoj Žižek, and the Kant-inspired anarcho-Marxism of Karatani Kojin.

48. V. I. Lenin's "A Retrograde Trend in Russian Social-Democracy," written in 1899, was first published in 1924 in *Proletarian Revolution*. Lenin, *Collected Works*, 255–85.

49. This delineation of crisis and disaster is taken from "Disaster, Crisis, Revolution," in Cazdyn, *Disastrous Consequences*.

50. Hardt and Negri, *Multitude*, 289.

51. Lauren Sedofsky, "Being by Numbers—Interview with Artists and Philosopher Alain Badiou," *Art Forum* (October 1994).

52. Žižek, *In Defense of Lost Causes*, 386.

53. Ibid., 404.

54. Karatani, *Transcritique*, 25.

55. Karatani develops this way of going beyond the trinity of capital-nation-state in two subsequent books, *Sekaikyowakoku e—shinon=neshon-kokka o koete* [Toward a World Republic: Transcending the Capitalist Nation-State] and *Sekai-shi no kozo* [The Structure of World History].

56. Karatani, *Sekai-shi no kozo*, x.

57. Karatani, *Transcritique*, 302–3.

58. "It seems to be easier for us today to imagine the thoroughgoing deterioration of the earth and of nature than the breakdown of late capitalism; perhaps that is due to some weakness in our imaginations." Jameson, *The Seeds of Time*, xii.

59. See Miyoshi's "Literary Elaborations" in *Trespasses*. "The death of humans is not the end of the planet. The earth, too, will ultimately end, like all other planets in the universe. But that has nothing to do with ecological problems. Planetary destruction requires a totally different kind of force. And as long as there is a physical earth, there will be other kinds of life on the planet—microbes, ants, rats, cockroaches, or whatever. And as long as life goes on, there will be other cycles of evolution. That is, after the end of humans, there will be life, other kinds of life that will evolve to produce their own civilizations. A new posthuman species will have its own life cycles. Of course, it may end up producing its own Bush-Cheney management. But that will take a great many millions, if not billions, of years. For now, a new kind of ecological studies will need to decide whether human extinction is worth thinking about. If it is, what about the cycles that will commence after ours is completed? At this time of very little hope all around, we can at least look forward to ongoing life on earth. And as long as we entertain any hope, we will manage to find the courage to keep trying." Miyoshi, *Trespasses*, 46–47.

60. Gretchen Morgenson and Don Van Natta Jr., "Paulson's Calls to Goldman Tested Ethics," *New York Times*, August 8, 2009.

61. Charles Abbott, "US Food Stamp List Tops 34 Million for First Time," *Reuters*, July 6, 2009.

62. Joe Bel Bruno, "Goldman Draws US Scrutiny," *Globe and Mail*, August 6, 2008. "Last week, New York Attorney-General Andrew Cuomo said Goldman paid 953 people bonuses of $1 million or more, including 212 who received $3 million or more. Overall, nine banks that received government aid money paid out bonuses of $32.6 billion last year—including more than $1 million apiece to nearly 5,000 employed, the report said."

63. Matt Taibbi, "Inside the Great American Bubble Machine," *Rolling Stone*, July 13, 2009.

64. *Wall Street Journal*, "A Tale of Two Bailouts," July 16, 2009; *Economist*, "Keeping up

with the Goldmans," July 15, 2009; Glenn Beck, "Why Goldman Sachs Is the Evil Empire!" *Glenn Beck*, Fox News Channel, July 15, 2009; Paul Krugman, "The Joy of Sachs," *New York Times*, July 16, 2009.

65. Stephen Gandel, "Goldman Sachs vs. Rolling Stone: A Wall Street Smackdown," *Time*, July 3, 2009.

66. Deirdre Bolton and Michael McKee, interview with Eliot Spitzer, "Spitzer Says Banks Made 'Bloody Fortune' with US Aid," Bloomberg Television, July 14, 2009.

67. Gretchen Morgenson and Don Van Natta Jr., "Paulson's Calls to Goldman Tested Ethics," *New York Times*, August 8, 2009.

68. Jameson, *The Geopolitical Aesthetic*, 3.

69. *The bcci Affair: A Report to the Committee on Foreign Relations, United States Senate, by Senator John Kerry and Senator Hank Brown*, December 1992, 102d Cong., 2d sess., Senate print 102–40, http://www.fas.org/irp/congress/1992_rpt/bcci/01exec.htm.

70. Brecht, *Happy End*.

71. Jodi Dean, "Feeling Sick," *I Cite* (blog), August 12, 2009, http://jdeanicite.typepad.com.

72. Mörtenböck and Mooshammer, *Networked Cultures*, 7.

73. Miyoshi gives his own account of this shift in "A Turn to the Planet: Literature, Diversity, and Totality," in *Trespasses*. See also my introduction to the book, "Trespasser: An Introduction to the Life and Work of Masao Miyoshi," xv–xxxiii.

74. Miyoshi, "Literary Elaborations," in *Trespasses*, 62.

75. Cazdyn, *The Flash of Capital: Film and Geopolitics in Japan*.

76. Shiva, *Monocultures of the Mind*.

77. Alain Badiou writes convincingly about the reactionary dimension of prevailing ethical principles in *Ethics: An Essay on the Understanding of Evil*.

78. This includes work for film and television. See the Internet Movie Database for a list of Miike's films.

79. Terry Eagleton writes, "There is a privileged, privatized hedonism about such [body] discourse, emerging as it does at just the historical point where certain less exotic forms of politics found themselves suffering a setback." *The Ideology of the Aesthetic*, 7.

80. Jacques Rancière, *The Future of the Image*, 138.

81. Andrea Hubert, "Film Stars of cctv," *Guardian*, October 13, 2007.

82. Since 1996, a group of anonymous New York–based performance artists known only as the Surveillance Camera Players have voiced their protest by staging plays directly in front of cctv cameras. They write, "It is all too often the case that— when faced with the alternative of identifying themselves and/or their audiences with the detectives who watch or the suspects who are watched—artists choose to identify with the detectives. We believe that such a choice strengthens several of the primary ideological supports for generalized surveillance: spectatorship, passivity, the allegedly superior intelligence of the watchers, the failure to understand or have sympathy for the psychologies of suspects, and the fetishization of technology and technological expertise." "Biting the Hands that Applaud Us," Surveillance Camera Players, October 9, 2007, http://www.notbored.org/faceless.html.

83. *Faceless: Chasing the Data Shadow*, Manu Luksch and Mukul Patel, http://www
.ambienttv.net/content/.

84. The Surveillance Camera Players have taken a hard line on Luksch's *Faceless*: "The
film does not demand or even advocate change; it simply declares that change has
already happened and we must 'question' it." "Biting the Hands that Applaud Us,"
Surveillance Camera Players, October 9, 2007, http://www.notbored.org/face
less.html.

85. Puchner, *Poetry of the Revolution*, 32, 79.

86. Ibid., 30.

Notes to Part 2: The Global Abyss

1. Maria Amuchastegui, "Farming It Out," *This Magazine*, May–June 2006.

2. Canadian Consulate General, letter to the author by M. Egan, Designated Immigration Officer, January 19, 2005.

3. Rosanna Tamburri, "Blessed Are the Benefactors," *University Affairs*, May 2004.

4. Ibid.

5. Nancy Fraser, "Transnationalizing the Public Sphere," *Republic Art*, March 2005,
http://www.republicart.net/disc/publicum/fraser01_en.htm.

6. "Bush Calls for Changes on Illegal Workers," John King, Ted Barrett, and Steve
Turnham, CNN.com, January 8, 2004, http://edition.cnn.com/2004/ALLPOLI
TICS/01/07/bush.immigration/.

7. Tirman, *The Maze of Fear*, 110.

8. Gerald T. Keusch, "Biomedical Globalization," *New England Journal of Medicine*
348, no. 24 (June 12, 2003): 2478.

9. Department of Health and Human Services, Office of Inspector General, Janet
Rehnquist, inspector general, *The Globalization of Clinical Trials: A Growing Challenge in Protecting Human Subjects*, September 2001, 3.

10. Indraji Basu, "South Asia's India's Clinical Trials and Tribulations," *Asia Times*, July
23, 2004.

11. China is also an important market. "Testing in China cuts the cost of clinical
trials—which can top $1 billion—by as much as a third," says Robert W. Pollard,
director of market research, Synovate Healthcare, China. "Go East, Big Pharma,
Drugmakers Are Expanding in China, But Patents Are Still a Worry," Bruce Ein-
horn, *Business Week* online, December 13, 2004, http://www.businessweek.com.

12. Clifford Krauss, "Going Global at a Small-Town Canadian Drugstore," *New York
Times*, March 5, 2005.

13. David Himmelstein, Deborah Thorne, Elizabeth Warren, and Steffie Woolhandler,
"Medical Bankruptcy in the United States, 2007: Result of a National Study,"
American Journal of Medicine 122, no. 8 (August 2009): 741–46.

14. Gina Shaw, "On a Quest for a Cure: International Patients Seek Top U.S. Hospitals
for Cancer Care," *Washington Diplomat*, 1997. About 66,000 foreign patients were
admitted to U.S. hospitals in 1996, while another 385,000 came for outpatient
treatments and physician consultations. That is up more than 50 percent from the
early 1990s.

15. Kathryn Schulz, "Did Antidepressants Depress Japan?" *New York Times*, August 23,

2004; Laurence J. Kirmayer, "Psychopharmacology in a Globalizing World: The Use of Antidepressants in Japan," *Transcultural Psychiatry* 39, no. 3 (September 2002): 295–322.

16. Schultz, "Did Antidepressants Depress Japan?"

17. One place to follow this shift is in the Canadian anticonsumerist magazine *Adbusters*.

18. Peter Boyle and Bernard Levin, eds., *World Cancer Report* (Switzerland: International Agency for Research on Cancer, 2008).

19. Arthur Caplan, "Indicting Big Pharma," *American Scientist*, January–February 2005.

20. Goozner, *The $800 Million Pill*. Goozner writes that the U.S. government (under the Reagan administration) spent about $10 billion on basic research to finally reach the AIDS triple cocktail by 1996. Private industry spent about $3 billion on licensing agreements and other funding for what, by 2004, became a $7 billion global market for the cocktail. *The $800 Million Pill*, 160.

21. Arnold Relman discusses taxpayer funding of research and development in his review of Daniel Vasella's *Magic Cancer Bullet: How a Tiny Orange Pill Is Rewriting Medical History*. Arnold Relman, "Magic Cancer Bullet: How a Tiny Orange Pill is Rewriting Medical History," *Journal of the American Medical Association* 290, no. 16 (2003): 2194–95.

22. Alex Berenson, "A Cancer Drug Shows Promise, at a Price That Many Can't Pay," *New York Times*, February 15, 2006.

23. Lisa Priest, "Dispute Blocks Cancer Drugs," *Globe and Mail*, June 19, 2006.

24. *Hilewitz v. Canada* (Minister of Citizenship and Immigration), 2002 FCT 844, IMM-4340–00, "Office of the Commissioner for Federal Judicial Affairs Canada," judgments of the supreme court of Canada.

25. This history is cited in the judgment of the Supreme Court of Canada, *Hilewitz v. Canada* (Minister of Citizenship and Immigration); *De Jong v. Canada* (Ministry of Citizenship and Immigration), [2005] 2 S.C.R. 706, 2005 SCC 57. An online version of this judgment can be found at http://csc.lexum.umontreal.ca/en.

26. Ibid.

27. Ibid.

28. Ibid.

29. Ibid.

30. Ibid.

31. Sontag, *Illness as Metaphor*, 87.

32. For this advertisement and others, see the website of the Canadian Psychiatric Research Foundation, http://cprf.ca/media/antiStigmaCampaign.html.

33. At the *Journal of Cell Biology*, the test has revealed the extensive manipulation of the photos. Since 2002, when the test was put in place, 25 percent of all accepted manuscripts have had one or more illustrations that were manipulated in ways that violate the journal's guidelines. See Nicholas Wade, "It May Look Authentic; Here's How to Tell It Isn't," *New York Times*, January 24, 2006. Much of this is driven by the stock we put in ways of proving medical conditions. With the success of the Human Genome Project, pictures of our genetic makeup have become the most secure form to prove our medical identities. In biometrics,

MRIs, and urine samples, the truth or falseness of our health is readily apparent for all to see; body scanners similarly search for infectious disease in the Shanghai airport.

34. *Chesters v. Canada* (Minister of Citizenship and Immigration), 2002 FCT 727, IMM-1316–97.

35. "More Staff Sent to Sri Lanka, Thailand to Fast-Track Visas for Relatives Canadian Press," *NewsWire*, January 5, 2005.

36. Judith Butler, "Is Kinship Always Already Heterosexual," *Differences: A Journal of Feminist Cultural Studies* 13, no. 1 (2002): 14–44, 16.

37. Ralf Michaels, "Same-Sex Marriage: Canada, Europe and the United States," *American Society of International Law Insights* (June 2003), http://www.asil.org/insigh 111.cfm.

38. Edelman, *No Future*.

39. Edelman strategically overidentifies with the right's attack on homosexuality—that homosexuality leads to the end of civilization and the death of the species. Instead of resisting this by stressing how responsible gays are and what loving parents they make, Edelman embraces the argument and calls for a queer ethics in which the queer inhabit the very negativity and danger that the right fears. Edelman refuses a politics that counterproductively articulates a better future.

40. Jacques Derrida, *Rogues*, trans. Pascale-Anne Brault and Michael Naas (Palo Alto: Stanford University Press, 2004), 25.

41. Figures change due to age, the type of match, and even the hospital where the transplant is performed.

42. *New Yorker*, May 9, 2005.

43. These comments were made by Wolfowitz in an address to delegates at an Asian security summit in Singapore in June 2003 and reported in George Wright, "Wolfowitz: Iraq War Was about Oil," *Guardian*, June 4, 2003. See also the correction to the article, http://www.guardian.co.uk/corrections/story/0,3604,971436,00 .html.

44. Agamben, *Homo Sacer*; Agamben, *State of Exception*.

45. Agamben, *Homo Sacer*, 147.

46. Ibid., 128.

47. Ibid., 132. Agamben reminds us that "one of the few rules to which the Nazis constantly adhered during the course of the 'Final Solution' was that Jews could be sent to the extermination camps only after they had been fully denationalized (stripped even of the residual citizenship left to them after the Nuremberg laws."

48. Žižek, *Welcome to the Desert of the Real*.

49. Agamben, *Homo Sacer*, 133.

50. I write "credited" because there is some disagreement about Trudeau's precise words and his position on the war resisters. Still, most remember Trudeau as being a strong critic of the Vietnam War and U.S. policy at the time.

51. By 2005, there were 5,133 U.S. troops missing from duty. Of these, 2,376 are sought by the army; 1,410 by the navy; 1,297 by the Marines; and 50 by the Air Force. Some have been missing for decades. Andrew Buncombe, "The Deserters: Awol Crisis Hits the US Forces," *Independent*, May 16, 2005.

1. "Goldman Sachs has developed a tradable index of life settlements, enabling investors to bet on whether people will live longer than expected or die sooner than planned. The index is similar to tradable stock market indices that allow investors to bet on the overall direction of the market without buying stocks." Jenny Anderson, "Wall Street Pursues Profit in Bundles of Life Insurance," *New York Times*, September 5, 2009.

2. Ibid.

3. Coventry First bought life insurance policies on behalf of AIG and Citigroup Inc. From Coventry's webpage. http://www.coventry.com/coventry-first/life-settle ments/life-settlement.asp

4. In 2006, for example, when the life settlements industry began to flourish and was posting more than $6 billion in annual revenue, Coventry First was accused of paying off a competitor to withdraw its higher bid on the policy of an eighty-year-old New York woman. "Spitzer Says Firm Defrauded Life Insurance Owners," Josh P. Hamilton and Hugh Son, Bloomberg.com, October 26, 2006, http://www.bloomberg.com/apps/news.

5. Donald MacKenzie, "End-of-the-World Trade," *London Review of Books* 30, no. 9 (May 8, 2008): 24–26.

6. Ibid.

7. Lacan, *The Seminars of Jacques Lacan: The Ethics of Psychoanalysis* (seminar 7), 184.

8. Latour, *We Have Never Been Modern*, 11.

9. We usually think about this in terms of language, the way the signified is stuffed by the signifier, as in Lacan's classic example of the two children coming into the station on the train. Lacan, *Ecrits*, 416.

10. Lacan, *Ecrits*, 123–58.

11. In fact, an emphasis on the late Lacan now dominates contemporary Lacanian thinking. We could argue that it is this late Lacanian Real that allows for such a diverse group of thinkers to claim Lacan as one of their more important interlocutors (Rancière, Badiou, Žižek, Miller, Copjec, Santner, Jameson, Karatani, Butler).

12. *Slavoj Žižek: The Reality of the Virtual*, DVD, directed by Ben Wright (London, United Kingdom: Ben Wright Film Productions, 2005).

13. This is also the way Jameson elaborates Žižek's notion of the parallax as a way of holding onto some aspect of the absolute and to resist a form of absolute relativism. See Jameson, *Valences of the Dialectic*, 62.

14. Freud, *Beyond the Pleasure Principle*.

15. Ibid., 77–78.

16. Lacan, *The Seminars of Jacques Lacan: The Four Fundamental Concepts of Psychoanalysis, 1964* (seminar 11), 257.

17. Lacan, *The Seminars of Jacques Lacan: The Ego in Freud's Theory and in the Technique of Psychoanalysis, 1954–1955* (seminar 2), 232.

18. Lacan, *The Seminars of Jacques Lacan: The Ethics of Psychoanalysis* (seminar 7), 243–83.

19. Žižek, *The Parallax View*, 62.

20. Ibid., 63.

21. Ibid.

22. *Slavoj Žižek: The Reality of the Virtual*, DVD, directed by Ben Wright (London, United Kingdom: Ben Wright Film Productions, 2005).

23. Nancy, *L'Intrus*, 3, 4.

24. Ibid., 8–9.

25. Ibid., 2.

26. Nancy, *Being Singular Plural*, 91.

27. Ibid., 92.

28. Agamben, *Homo Sacer*, 159.

29. Ibid., 18.

30. Foucault, *History of Sexuality*, 141–42.

31. Agamben, *Homo Sacer*, 10.

32. Agamben, *Means without Ends*, 114–15.

33. Agamben, *The Open*, 82.

34. Agamben, *Homo Sacer*, 164.

35. Ibid., 162.

36. Ibid., 165.

37. Lock, *Twice Dead*, 4.

38. Ibid., 11.

39. Ibid., 377.

40. Shimazonon Susumu presents an overview of the program on the program's website.

41. Herbert Marcuse, "The Ideology of Death," in Feifel, *The Meaning of Death*, 67.

42. Ibid., 74.

43. Bloch, *The Principle of Hope*, 1180.

44. Ibid., 1182.

45. Jameson, *Marxism and Form*, 183.

46. Ibid.

47. Jameson, *Valences of the Dialectic*, 26.

48. Jung, "Psychological Commentary," in *The Tibetan Book of the Dead*, xli. "The only initiation process that is alive today and practiced in the west is the analysis of the unconscious used by doctors for therapeutic purposes."

49. See Geoffrey Blowers, "Freud's Deshi: The Coming of Psychoanalysis to Japan," *Journal of the History of the Behavioral Sciences* 33, no. 2 (1998): 115–26.

50. Freud, *Civilization and Its Discontents*, 62.

51. Marcuse, *Eros and Civilization*, 35.

52. Ibid., xxv.

53. On Marcuse's "Political Preface" (1966) in *Eros and Civilization*, as well as the relation between Marcuse and Lacan, see the very interesting essay by Daniel Cho, "Thanatos and Civilization: Lacan, Marcuse, and the Death Drive," *Policy Futures in Education* 4, no. 1 (2006).

Agamben, Giorgio. *Language and Death: The Place of Negativity*. Translated by Karen Pincus and Michael Hardt. Minneapolis: University of Minnesota Press, 2006.

——. *State of Exception*. Translated by Kevin Attell. Chicago: University of Chicago Press, 2005.

——. *The Open*. Translated by Kevin Attell. Palo Alto: Stanford University Press, 2003.

——. *Means without Ends*. Translated by Vincenzo Binetti and Casare Casarino. Minneapolis: University of Minnesota Press, 2000.

——. *Homo Sacer: Sovereign Power and Bare Life*. Translated by Daniel Heller-Roazen. Palo Alto: Stanford University Press, 1998.

Badiou, Alain. *Metapolitics*. London: Verso, 2005.

——. *Ethics: An Essay on the Understanding of Evil*. Translated by Peter Hallward. London: Verso, 2001.

Bloch, Ernst. *The Principle of Hope*. Vol. 3. Translated by Neville Plaice, Stephen Plaice, and Paul Knight. Boston: MIT Press, 1995.

Brecht, Bertolt. *Happy End: A Melodrama with Songs*. New York: Samuel French, 1929.

Cazdyn, Eric. "Disaster, Crisis, Revolution." *South Atlantic Quarterly* 106, no. 4 (2007): 647–62.

——. *The Flash of Capital: Film and Geopolitics in Japan*. Durham: Duke University Press, 2002.

Cousins, Norman. *Anatomy of an Illness as Perceived by the Patient*. New York: Norton, 1979.

Dick, Philip K. *The Minority Report*. New York: Pantheon, 2002.

Dienst, Richard. *Still Life in Real Time*. Durham: Duke University Press, 1994.

Eagleton, Terry. *The Ideology of the Aesthetic*. London: Blackwell, 1990.

Edelman, Lee. *No Future: Queer Theory and the Death Drive*. Durham: Duke University Press, 2004.

Evans-Wentz, W. Y., ed. *The Tibetan Book of the Dead*. London: Oxford University Press, 1960.

Feifel, Herman. *The Meaning of Death*. New York: McGraw-Hill, 1959.

Fink, Bruce. *A Clinical Introduction to Lacanian Psychoanalysis: Theory and Technique*. Cambridge: Harvard University Press, 1997.

Foucault, Michel. *History of Sexuality*. Vol. 1. Translated by Robert Hurley. New York: Vintage, 1990.

Freud, Sigmund. *Interpretation of Dreams*. Translated by Joyce Crick. Oxford: Oxford University Press, 2008.

——. *Wild Analysis*. Edited by Adam Phillips. New York: Penguin, 2002.

——. *The Complete Letters of Sigmund Freud to Wilhelm Fliess, 1887–1904*. Translated by Jeffrey Moussaieff Masson. Cambridge, Mass.: Belknap, 1986.

——. *Beyond the Pleasure Principle*. Translated by James Strachey. New York: Norton, 1961.

Freud, Sigmund, and Joseph Breuer. *Studies in Hysteria*. Translated by Nicola Luckburst. New York: Penguin, 2004.

Goozner, Merrill. *The $800 Million Pill: The Truth behind the Cost of New Drugs*. Berkeley: University of California, 2005.

Hanks, Geoffrey, Nathan I. Cherny, Nicholas A. Christakis, Marie Fallon, Stein Kaasa, and Russell K. Portenoy, eds. *Oxford Textbook of Palliative Care*. Oxford: Oxford University Press, 2009.

Hardt, Michael, and Antonio Negri. *Multitude*. New York: Penguin, 2005.

Jameson, Fredric. *Valences of the Dialectic*. London: Verso, 2009.

——. *The Seeds of Time*. New York: Columbia University Press, 1996.

——. *The Geopolitical Aesthetic: Cinema and Space in the World System*. Bloomington: Indiana University Press, 1992.

——. *Marxism and Form*. Princeton: Princeton University Press, 1974.

Jopling, David. *Talking Cures and Placebo Effects*. London: Oxford University Press, 2008.

Karatani Kojin. *Sekai-shi no kozo* [The Structure of World History]. Tokyo: Iwanami Shoten, 2010.

——. *Sekaikyowakoku e—shinon=neshon-kokka o koete* [Toward a World Republic: Transcending the Capitalist Nation-State]. Tokyo: Iwanami Shoten, 2006.

——. *Transcritique: On Kant and Marx*. Translated by Sabu Kohso. Cambridge: MIT Press, 2005.

Kerr, Clark. *The Uses of the University*. Cambridge: Harvard University Press, 2001.

Lacan, Jacques. *Ecrits*. Translated by Bruce Fink. New York: Norton, 2007.

——. *The Seminars of Jacques Lacan: The Four Fundamental Concepts of Psychoanalysis, 1964*. Book 11. Edited by Jacques Alain-Miller. Translated by Alan Sheridan. New York: Norton, 1998.

——. *The Seminars of Jacques Lacan: The Ethics of Psychoanalysis*. Book 7. Edited by Jacques Alain-Miller. Translated by Dennis Porter. New York: Norton, 1997.

————. *The Seminars of Jacques Lacan: The Ego in Freud's Theory and in the Technique of Psychoanalysis, 1954–1955.* Book 2. Edited by Jacques Alain-Miller. Translated by Sylvana Tomaselli. New York: Norton, 1991.

Latour, Bruno. *We Have Never Been Modern.* Translated by Catherine Porter. Cambridge: Harvard University Press, 1993.

Law, Jacky. *Big Pharma: How the World's Biggest Drug Companies Control Illness.* New York: Carroll and Graf, 2006.

Le Fanu, James. *The Rise and Fall of Modern Medicine.* London: Abacus, 2001.

Lenin, V. I. *Collected Works.* Vol. 4. 4th English edition Moscow: Progress, 1972.

Lock, Margaret. *Twice Dead: Organ Transplants and the Reinvention of Death.* Los Angeles: University of California Press, 2001.

Ludmerer, Kenneth. *Time to Heal: American Medical Education from the Turn of the Century to the Era of Managed Care.* London: Oxford University Press, 2005.

Marcuse, Herbert. *Eros and Civilization.* New York: Vintage, 1966.

Marx, Karl. *Capital: A Critique of Political Economy.* Vol. 1. Translated by Ben Fowkes. New York: Vintage Books, 1977.

Miyoshi, Masao. *Trespasses: Selected Writings.* Durham: Duke University Press, 2010.

Moore, Adam D. *Information Ethics: Privacy, Property and Power.* Seattle: University of Washington, 2005.

Mörtenböck, Peter, and Helge Mooshammer. *Networked Cultures: Parallel Architectures and the Politics of Space.* Rotterdam: NAI, 2008.

Nancy, Jean-Luc. *L'Intrus.* Translated by Susan Hanson. Detroit: Michigan State University Press, 2002.

————. *Being Singular Plural.* Translated by Robert Richardson and Anne O'Byrne. Palo Alto: Stanford University Press, 2000.

Puchner, Martin. *Poetry of the Revolution: Marx, Manifestos, and the Avant-Gardes.* Princeton: Princeton University Press, 2005.

Ratledge, Colin, and Bjørn Kristiansen. *Basic Biotechnology.* Cambridge: Cambridge University Press, 2001.

Roudinesco, Élisabeth. *Why Psychoanalysis?* New York: Columbia University Press, 2003.

————. *Jacques Lacan.* Translated by Barbara Bray. New York: Columbia University Press, 1999.

Schneiderman, Stuart. *Lacan: The Death of an Intellectual Hero.* Cambridge: Harvard University Press, 1984.

Schur, Max. *Freud: Living and Dying.* London: Hogarth, 1972.

Shiva, Vandana. *Monocultures of the Mind: Perspectives on Biodiversity and Biotechnology.* London: Zed, 1993.

Sontag, Susan. *Illness as Metaphor.* New York: Farrar, Straus, Giroux, 1988.

Tirman, John, ed. *The Maze of Fear: Security and Migration after 9/11.* New York: New Press, 2004.

Uno Kozo. *Marukusu keizaigaku genriron no kenkyu* [Research on Marxist Economic Principles]. Tokyo: Iwanami Shoten, 1959.

Vasella, Daniel. *Magic Cancer Bullet: How a Tiny Orange Pill Is Rewriting Medical History*. New York: Harper Business, 2003.

Virilio, Paul. *The Vision Machine*. Bloomington: Indiana University Press, 1994.

Žižek, Slavoj. *In Defense of Lost Causes*. London: Verso, 2009.

——. *The Parallax View*. Cambridge: MIT Press, 2006.

——. *Welcome to the Desert of the Real*. London: Verso, 2002.

——. *The Sublime Object of Ideology*. London: Verso, 1989.

body: commodity of, 85–86; in film, 83–86; reaction to organ transplant, 178–79; technologies of, 76

Brecht, Bertolt, 65

Bristol-Myers Squibb, 124

Buddhism, 184, 190–92

Buettner, Jennifer, 25

Bush, George W., 63, 121, 157; on immigration, 110–11, 113; on preemptive war, 130–31; on same-sex marriage, 138–39

Bush Doctrine, 131

Butler, Judith, 137, 141

Caché (Haneke), 37–42, 207 n. 37

Canadian health care system: fragility of, 107–8; government spending and costs for, 20–21, 115, 136; health insurance and, 101; refusal to treat patients, 133

Canadian immigration: Canada-U.S.-Mexico immigration debate, 110; demographic changes in, 136; discrimination in, 128; family class influence on, 135–40; immigrant workers, 107–8; Immigration Acts, 127; international adoption policies of, 139–40; law of, 126–27; medical inadmissibility cases, 125–28; new immigrants, 136; permanent residency application of, 129, 135–36, 151; policy of, 115, 132, 139, 159; refugee status and, 156–57; transnational citizenship and, 112–13

Canadian Patented Medicine Prices Review Board, 124

Canadian Psychiatric Research Foundation, 133

cancer: capitalism and, 58–59; cure for, 18, 59, 101, 128; detection technologies for, 18–19, 130; diagnosis of, 168–69, 198–99; drug costs and, 124, 206 n. 4; Hirohito's death from, 83; metaphors for, 130; physiological autonomy of, 167; preemptive medicine and, 131–32; prostate, 18–19, 205 n. 2; WHO cancer rate estimates, 119. *See also* Chronic Myelogenous Leukemia (CML)

Capital (Marx), 33–34, 58

capitalism: anticapitalism principle, 57; cancer and, 58–59; the chronic and, 5, 68; commodities and, 2, 33–34, 67, 94, 139, 196; conspiracy and, 63–64; crime of, 65, 78–79; crisis and, 2–3, 62, 94, 140, 151, 196; and death insurance, 188; desires for an alternative to, 204; as an economic force, 61–62, 64; end of, 61–62, 161, 203; exploitation, 2; and family normatives, 139, 141–42; globalization as a stage of, 77, 115; hypocrisy of consumer, 80; ideology of, 49, 79, 81, 193, 203; and labor, 2, 42; late capitalism, 163, 209 n. 58; mode of production for, 2, 33, 58, 61, 79, 115; morality and, 82–83; neoliberal, 109; nonmoralizing critique of, 195–200; pharmaceutical corporations and, 155; representation and misrepresentation of, 34–35; reproduction of, 53; revolutionary Marxism and, 47–48, 208 n. 46; system of, 34–35, 67; taboo in America, 67; transgressive acts, 52, 94; vulgar capitalism, 48–49, 202. *See also* global capitalism

Charter of Rights and Freedom (Canada), 135

Chesters, Angela, 135

Chinese medicine, 22, 199

chronic: concept of, 5; culture, 69–71; disease, 20–22, 206 n. 4; temporality, 70, 72, 83; the terminal and, 4, 6, 14, 44, 46, 72, 204. *See also* chronic time; new chronic

Chronic Myelogenous Leukemia (CML): bone marrow transplant, 147; chromosomal mutation, 145–46; first-line treatment, 119–20; Gleevec costs and accessibility, 120–25, 148; Gleevec treatment, 100–101, 129, 146; Internet support group, 119–22; remission, 128, 146

chronic time, 6, 36, 41, 195; marriage and, 140; totality vs., 174

Civilization and Its Discontents (Freud), 154, 192–93

Clinton, Bill, 138

Closed Caption Television (CCTV), 91–93, 210 n. 82

Cold War: crisis and disaster and, 55; jokes, 96; political categories, 49; revolutionary discourses of, 59–60. *See also* post-Cold War era

commodities: capitalism and, 2, 33–34, 67, 94, 139, 196; consumerism and, 35, 52, 148; death and, 161–63

commodity culture, 20, 53, 79, 151

commodity production, 62, 68, 105, 115–16, 155

communism, 50, 96, 162, 193

Complex Chronic Disease (CCD), 20–21

conspiracy: capitalism and, 63–64; films, 64–68; jokes, 80; mapping the system of, 66–67

Cousins, Norman, 24–25

Coventry First, 214 n. 3

crime: of capitalism, 65, 78–79; in film, 78, 82, 95; surveillance of, 91, 94–95

crisis: AIDS, 118; cancer and, 131–32; capitalism and, 2–3, 62, 94, 140, 151, 196, 199; countries in, 131; culture in, 69, 71; cyberpunk and, 76; disaster difference and, 53–55; global catastrophe and, 154; of the globalized nation-state, 156; images representing, 87; manufacturing, 69; meaning of, 3–4, 46–47; new chronic and, 44; of political representation, 150; short-term vs. long-term, 3, 86, 198; system in, 1; time and, 3–4, 11, 75

cultural analysis, 15, 72–73

cultural manifestos, 72

culture: commodity and, 20, 53, 79, 151; crisis of, 69, 71; ideological shifts of, 86; national, 50, 103; network art and, 70; popular, 44, 119, 140, 195; prescriptive of, 70–71; reality of, 69–71; revolutionary, 6, 71–73, 75; stereotypes and, 118; transplants in Japan and, 183

cure: border crossings in search of, 117; for cancer, 18, 59, 101, 128; desire for, 6, 15, 27, 204; disease management and, 9, 18–20, 58–59, 128–30, 199–200; profits from, 118; revolution and, 7, 15, 47, 56, 60–61; talking cures, 26–27; time and, 28, 36

Cure (Kurosawa), 82, 84

cyberpunk films, 75–77, 83

cyborg, 75–77, 83

Dean, Jodi, 66–67

death: "always already dead" concept, 163, 165; "always dead" concept, 163–65; autonomy of, 7, 185; chronic time and, 5–6, 193; comic book deaths, 194–95; commodification of, 161–63; discontinuity of, 186–87; euthanasia, 5, 161, 186; in film, 41–42, 76, 178, 194; of god, 163–65; human experience of, 158; the Imaginary and, 166–67; insurance, 160, 163, 188; language and, 182; life binary and, 180, 188; living dead, 8, 177–78, 180, 187; nonmoralizing critique of capitalism and, 195, 198–200; ontologies in Japan, 183–85; over-comatose condition and, 182–83; palliative care, 44–46, 184–85; radical change and, 7; the Real and, 173; right to die, 163–64; the Symbolic and, 167; *The Tibetan Book of the Dead*, 189–91; time and, 8, 14, 194; "twice dead" concept, 181, 183; "undead" concept, 8, 175–78, 180, 187; utopia and, 163, 185–87. *See also* already dead; life and death relationship

death drive, 141, 174–77, 192

De Jong v. Canada, 125, 127, 212 n. 25

democracy, 49, 60, 154, 181; capitalism and, 61, 132; representational, 142–44

depression: antidepressants, 24, 118, 205 n. 3; critique on, 24; in Japan, 118–19

Derrida, Jacques, 165

desire: to cure, 6, 15, 27, 204; drive and, 175–76, 191; from the image, 87–89, 91, 95; for representation, 87; unconscious, 29, 95, 172–73, 195

detection technologies, 18–19, 119, 130

Dewey, John, 52

diagnosis: of leukemia, 9–10, 100, 107, 129, 150; limitations of medical imaging and, 168–73; overdiagnosis, 18–19

Diagnostic and Statistical Manual of Mental Disorders (DSM), 24, 42

Dick, Philip K., 77, 95

disaster: crisis difference and, 53–55; uneven effects of, 14

disease management, 6, 21–22, 130, 132, 160; palliative care and, 46, 208 n. 41;

disease management (*cont.*)
    pharmaceutical interests in, 19, 26; role
    of the cure in, 9, 11, 19, 58–59
DNA research, 134
DreamWorks, 77–78
drive, 8, 174–77
drugs: accessibility of, 120–25; anti-
    depressants, 24, 118, 205 n. 3; anti-
    retrovirals, 13, 23, 81, 102, 212 n. 20; Ava-
    stin, 124; Erbitux, 124; generic drugs,
    81, 123, 158; Letrozole, 116; Paxil, 118;
    Prozac, 24, 118; psychotropics, 27, 205
    n. 3; trials, 116; Vioxx, 24, 206 n. 12. *See
    also* Gleevec

Eagleton, Terry, 85, 210 n. 79
ecology, 53
economics, 2, 128, 199, 203; bailouts, 7, 63,
    66–67; bioeconomic, 152–55; class
    applicants and medical inadmissibility,
    135–36, 140, 143, 156; critique on cap-
    italism and, 56–58; determinism, 49–
    52; force, 62, 64; global political prob-
    lems and, 113, 115; logic, 48, 86, 152, 155;
    pharmaceutical corporations and, 123–
    24; political realities and, 103, 121; sys-
    tem, 80, 112, 150
economic crisis (2008), 2–3, 7, 159, 161
economism, 50–51
Economist, 62
Edelman, Lee, 141–42, 213 n. 39
ego, 166, 174
Eli Lilly, 24, 118
environmental studies, 74
Erbitux, 124
eros, 192–93
Eros and Civilization (Marcuse), 185, 193
eugenics, 86, 133, 153
euthanasia, 5, 161, 186

Faceless (Luksch), 72, 76, 91–95, 97, 211
    n. 84
family: capitalism and, 141–42; class priv-
    ileges in immigration policy, 136–37,
    139, 143; support for the disabled in
    immigration law, 126–28
film: amnesia themes in, 72, 76–83;
    chronic time in, 40–41; conspiracy

themes in, 64–68; cyberpunk, 75–77,
    83; dystopian themes in, 95–97; mar-
    riages of convenience in, 140; portrayal
    of the body in, 83–86; presumed death
    in, 194; psychoanalytical themes in, 2,
    36–43; representation of time in, 43–
    44, 89–90, 97–98; surveillance images
    used in, 91–95; undead themes in, 176–
    81; viewership of, 81; zombies in, 201–4
film stock, 88–90
Fink, Bruce, 207 n. 26
Food and Drug Administration (FDA),
    116, 120, 123
food stamp program, 62
Fortune 500, 23
Foucault, Michel, 145, 165, 181
Fox, Vincente, 110–11, 113
Freud, Sigmund: on cures, 27–29; death
    drive principle, 175; on dreams, 207
    n. 29; on eros, 192–93; film and, 78;
    fraud in psychoanalysis and, 29–30;
    Jung's criticism of, 191–92; letter to
    Marie Bonaparte, 5; the pleasure prin-
    ciple and, 174; reader of Freud text, 35;
    on repression of instincts, 154;
    response to Romain Rolland, 192; self-
    analysis struggles, 31–32; on termi-
    nality of psychoanalytic treatment, 30;
    on trauma, 177; writing style of, 32–33
future, the: capitalism and, 141–42, 197;
    death and, 162–63, 165, 187, 189, 200; in
    film, 92–97; illness management and,
    9, 19, 22, 160; limitations of medical
    imaging and, 19, 168–69, 172–73; the
    present and, 4–9, 14, 46, 74, 83, 98;
    preventative treatment and, 30, 132,
    149; reproductive futurity, 141
Future of the Image, The (Rancière), 87

Genentech, 124
General Electric (GE), 149
geopolitical system, 60, 76, 152, 153, 156
GlaxoSmithKline, 118
Gleevec: advertising for, 147–48; costs
    and accessibility, 120–25, 150; treat-
    ment, 100–101, 129, 146
global abyss: the already dead and, 200;
    concept of, 7–10, 103; the foreigner and,

medical inadmissibility: case studies, 125–28; exceptions, 156

medical metaphors, 14

medical screening, 18–19, 205 n. 2

medical tourism, 116–17

mental illness, 24, 26, 35; in Japan, 118; physical illness vs., 132–34

Merck, 206 n. 12

Mexican health care system, 107–8

Mexico-U.S. border, 110–11, 113

Michael Clayton (Gilroy), 194

migrant workers, 105–6

military logic, 4

military state, 17–18

Minority Report (Spielberg), 95

Miyoshi, Masao, 73–74, 209 n. 59

modernization theory, 73, 105–6

modern power, 152–53

Mooshammer, Helge, 70–71

Mörtenböck, Peter, 70–71

Motion Picture Association of America, 80

Multitude (Hardt & Negri), 56–59

Munchausen syndrome, 132

Nancy, Jean-Luc, 8, 178–80, 185, 187

narratives: comic book, 194–95; cyborg, 75–76; in film, 38–40, 92, 94–97, 140; illness, 68; Japanese prose, 73–74

Natco Pharmaceuticals, 123

National Institutes of Health, 116

nationalism, 11, 51, 115, 154

nation-state, 109, 115, 153; global integration of power and, 64–65; globalization and transformation of, 77, 106, 142, 156, 158; global system and, 7, 103; sovereignty of, 76, 106, 110, 138

nativity, 153

natural disasters, 53–54

Negri, Antonio, 56–59, 208 n. 47

Networked Cultures (Mörtenböck & Mooshammer), 70–71

neurosis, 32, 175, 201

new chronic: concept of, 6–11, 13; as dominant temporality, 49, 76, 140; in film, 92; of global capitalism, 195; as illness management strategy, 132; logic of time, 17; palliative care and, 44–46

New Yorker, 147–48

New York Times, 62–63, 212 n. 33

9/11, 101, 113, 154

"nocebo effect," 25

No Logo (Klein), 52

North American Free Trade Agreement (NAFTA), 100, 111

Obama, Barack, 66, 144

Obecalp, 25

Oedipus, 175

Oedipus complex, 191

Oklahoma City bombing, 113–14

online pharmacies, 116

Ontario Health Insurance Plan (OHIP), 101, 129

Ontario Trillium Program, 124–25

organ trade, 117

organ transplantation, 178–79, 183

Oshima Nagisa, 85

overcomatose, 182–83

palliative care, 44; end-of-life care in Japan, 184–85; marginalization of, 45–46, 199

Paulson, Henry, 63

Paxil, 118

Paycheck (Woo), 78

Pfizer, 206 n. 4

pharmaceutical companies: academic partnerships with, 122–23, 155; advertising and marketing, 122; capitalist expansion of, 22–23; clinical trials, 116, 120; control over medical practice, 70; drug pricing practices, 122–24, 155; Eli Lilly, 24, 118; financial interests in disease management, 19–20, 26; Genentech, 124; generic drugs, 81, 123, 150; GlaxoSmithKline, 118; Merck, 206 n. 12; Natco Pharmaceuticals, 123; online pharmacies, 116; patents, 23, 47; Pfizer, 206 n. 4; placebos and the pharmaceutical model, 25–26; profits and funding, 23–24, 27, 118, 148

pharmacotherapy, 24

Philosophy of Right (Hegel), 58

placebos, 24–26

Place of Solitude, 71

voters/voting, 143–44
vulgar capitalism, 48–49, 202

Wall Street, 3, 62–63, 66, 160
Wall Street Journal, 62
Watson, James, 134
Wolfowitz, Paul, 155–56, 213 n. 43
World Bank, 117, 143
World Health Organization (WHO), 45,
  119

Yabe Yaekichi, 192
Yoshida Kiju, 88–91, 97

Zapatistas, 189–90
Žižek, Slavoj: on absolute relativism, 214
  n. 13; on functional health, 59; joke
  told by, 17; on Lacan's three orders, 171;
  on revolution, 56–57; on the undead,
  8, 175–77, 180, 185, 187
zombies, 176, 201–4

Eric Cazdyn is professor of critical and cultural theory in the Centre
for Comparative Literature and the Department of East Asian studies
at the University of Toronto. He is the author of *After Globalization*
(with Imre Szeman, 2011) and *The Flash of Capital: Film and Geopolitics
in Japan* (2002), and editor of *Trespasses: Selected Writings of Masao
Miyoshi* (2010) and *Disastrous Consequences* (2007).

Library of Congress Cataloging-in-Publication Data
Cazdyn, Eric M.
The already dead : the new time of politics, culture, and illness /
Eric Cazdyn.
p. cm.
Includes bibliographical references and index.
ISBN 978-0-8223-5203-7 (cloth : alk. paper)
ISBN 978-0-8223-5228-0 (pbk. : alk. paper)
1. Capitalism—Social aspects.
2. Globalization—Social aspects.
3. Globalization—Health aspects.
4. Emigration and immigration—Social aspects.
5. Emigration and immigration—Health aspects.
I. Title.
HB501.C39 2012
306.3—dc23   2011048242